MINING IN
CORNWALL

MINING IN CORNWALL

VOLUME 8
CAMBORNE TO REDRUTH

L.J. Bullen

The History Press

In memory of my great-uncle, the late Clarence King. He carried
out a geological survey in the state of Arizona for the US Federal
Government in the 1870s. Sometime later in that same state he
was instrumental in exposing the most sensational attempted
mine 'salting' (fraud) episode in American history.

First published by Landmark Publishing Ltd in 2006
This edition published 2013

The History Press
The Mill, Brimscombe Port
Stroud, Gloucestershire, GL5 2QG
www.thehistorypress.co.uk

British Library Cataloguing in Publication Data.
A catalogue record for this book is available from the British Library.

ISBN 978 0 7524 9310 7

Typesetting and origination by The History Press
Printed in Great Britain

Contents

Acknowledgements

The production of this book has been greatly facilitated by the following friends and members of my family – Brian Errington, David Gill, Peter Gilson of the Royal Cornwall Polytechnic Society, John Treloar, my wife Margaret, my daughter Anne Smith and my son-in-law Robert Smith. I must also record my thanks to The History Press for their support and encouragement.

Introduction

This volume reviews a selected group from the many mines between Camborne and Redruth. These include enterprises both large and small which were a part of the most intensively mined area in the county.

In this book I have portrayed a group of mines within the Central Mining District of Cornwall. These include the famous Dolcoath mine.

It is well known that Cornwall is acknowledged internationally as the home of hard-rock mining and the cradle of the steam engine. The rich mineral deposits with which the county is historically endowed created the desire to seek the knowledge and skills to discover and mine them.

Many other industries, particularly engineering, evolved because of the remoteness of the peninsula. Although transport was available by sea, a rail link with the rest of the United Kingdom came as late as 1859. Hence the emergence of companies whose fame spread to every mining area of the globe. As a result, a number of these outlived the indigenous mining industry and created employment and stability until the last quarter of the twentieth century by exporting overseas. Their final demise only reflects the fact that Great Britain is no longer a major manufacturing country.

It is felt that this second edition of Volume 8 of the 'Mining in Cornwall' series by The History Press will be welcomed by all who have an interest in the remarkable history of metalliferous mining within the county of Cornwall.

L.J. Bullen
Camborne, Cornwall
July 2013

1

Carn Camborne and Dolcoath Mines

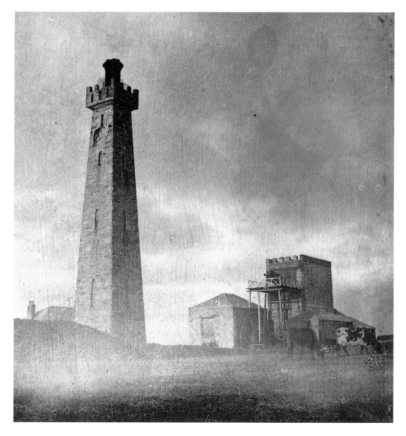

Carn Camborne mine. From left to right: the unusual stack, copper crusher and the 22in cylinder rotary engine house. This engine drove the pumps and crusher and also wound from the shaft, the headgear of which can be seen on the extreme right-hand edge of the print. The roof and one chimney of the counthouse are visible behind the stack. This sett was acquired by Dolcoath mine in 1898. (*c.* 1880)

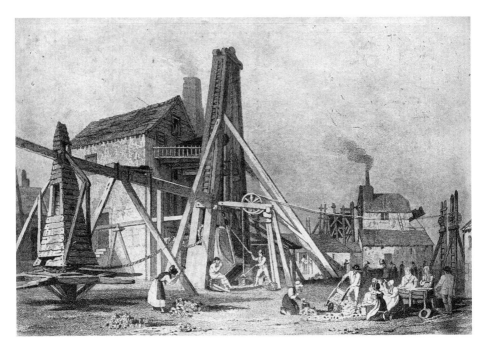

An interesting old engraving with much artistic licence showing New Sump shaft pumping engine on the left. To the right is what appears to be a rotary engine with a balance weight on the nose of the beam. This could be representing one of the Sims type single-acting engines, but the artist has omitted the sweep rod. (Early nineteenth century)

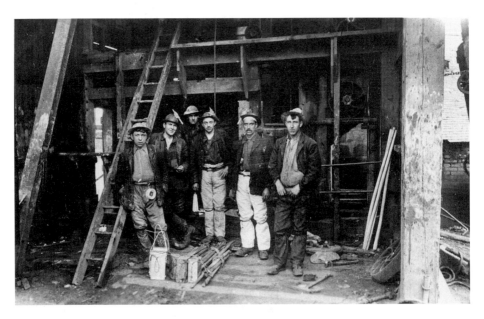

A group of miners at New Sump shaft in the 1890s.

Taken at the
388 fathom
level showing
very heavy
timbering.
(1890s)
*(Photograph:
J.C. Burrow
FRPS)*

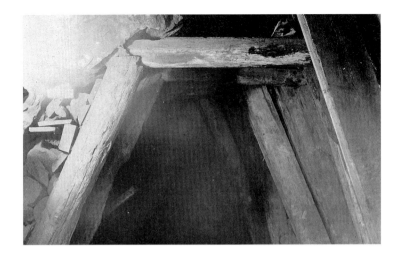

A remarkable
photograph
showing miners
sitting on stull
timber above a
huge gunnies.
(1890s)
*(Photograph:
J.C. Burrow FRPS)*

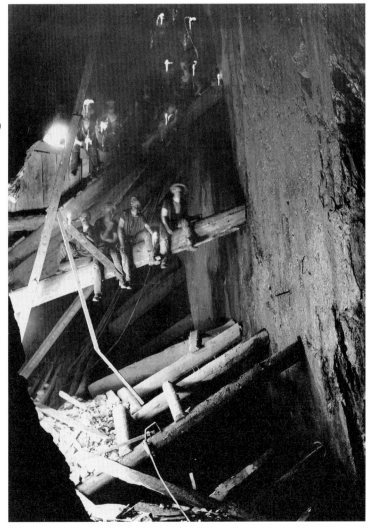

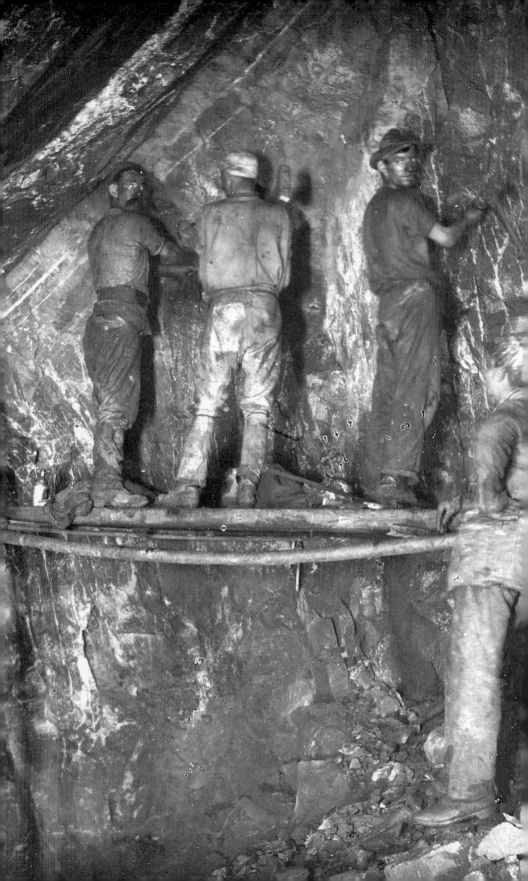

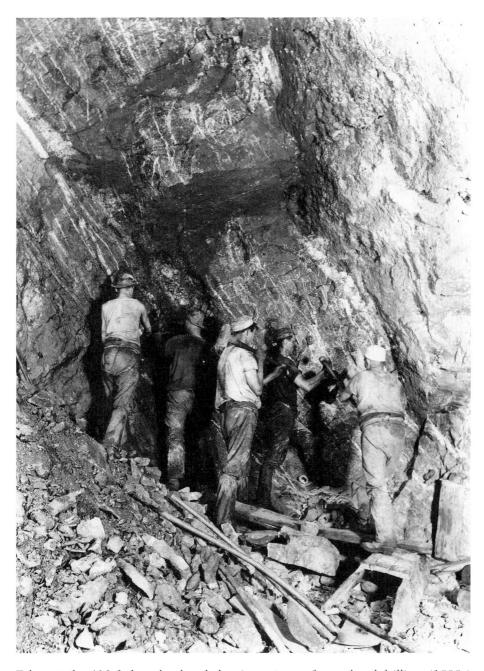

Taken at the 412 fathom level and showing a team of men hand-drilling. (1890s)
(Photograph: J.C. Burrow FRPS)

Opposite: A team of four men drilling (hand labour) on a conventional three-leg stage. They are breast stoping. The figure leaning on the staging is a mine captain. (1890s) *(Photograph: J.C. Burrow FRPS)*

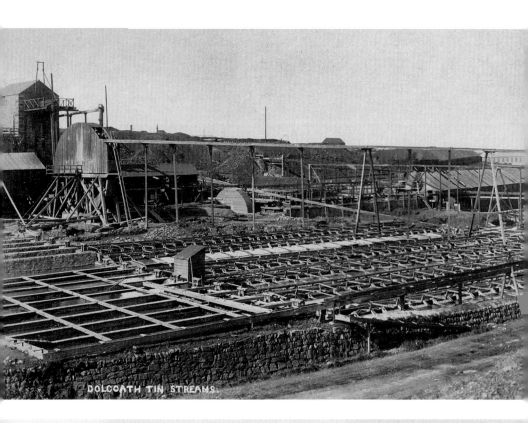

DOLCOATH TIN STREAMS.

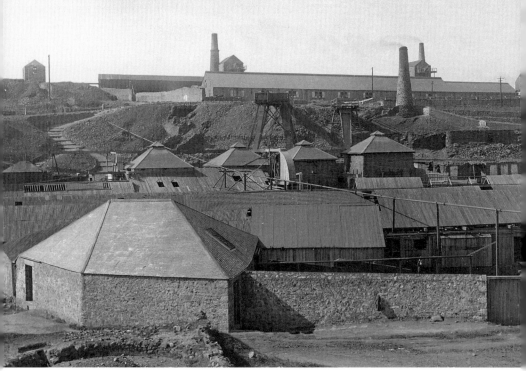

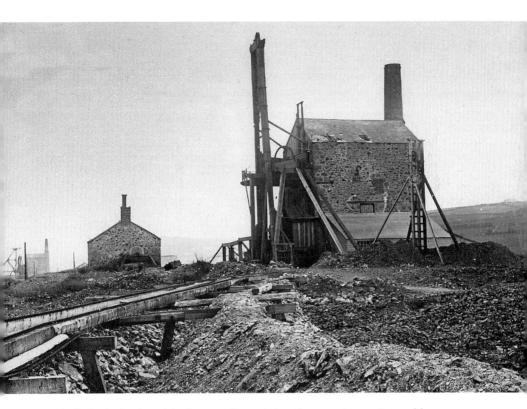

Straypark shaft. At the time this photograph was taken the pumping engine and house were in a parlous state. Early in the twentieth century the house was rebuilt, except for the stack, and Holman Brothers of Camborne manufactured a new 65in cylinder engine. Only the bob of the old engine was retained. (*c.* 1898)

Opposite top: A view from the eastern side of the Tuckingmill Valley. In the floor of the valley are settling pits and a large number of rag frames. The old stamps engine is on the left with a large dipper wheel, which is lifting the pulp to the launder seen crossing the valley to the treatment plant below the Californian stamps out of shot on the right. The dipper wheel was powered by the stamps engine. (1890s)

Opposite bottom: A further view of the Tuckingmill Valley. At the top are two Cornish stamps engines and the buildings containing the dressing plant. On the right is the structure which housed the early Elmore Oil Flotation Plant, which was installed to recover sulphide ores from some of the shallower workings of the mine. The building on the left skyline is the beam winder house of New East shaft. The two structures in the centre of the scene are at the end of tramway gantries, which carried the ore from the table house to the four water-powered calciners below. (1890s)

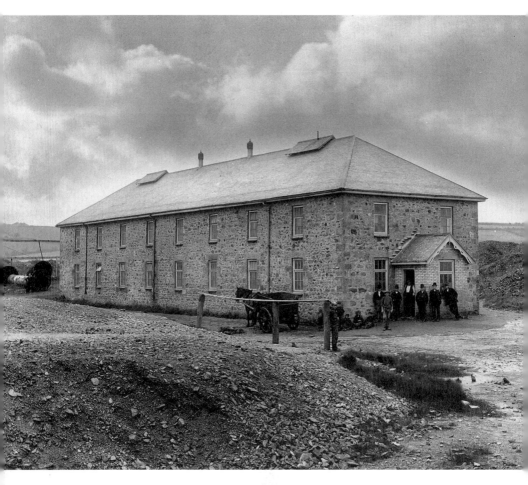

The mine dry when new in 1888. This change house, always known as a dry in Cornwall, was built near Wheal Harriett shaft and was state of the art at the time. Changing facilities on most mines were rudimentary to say the least.

Opposite: Wheal Harriett shaft showing a part of the gearwork of the 65in pumping engine which had recently ceased working. It will be noticed that the engine is greased over. (1921) (*Photograph: W.A. Michell*)

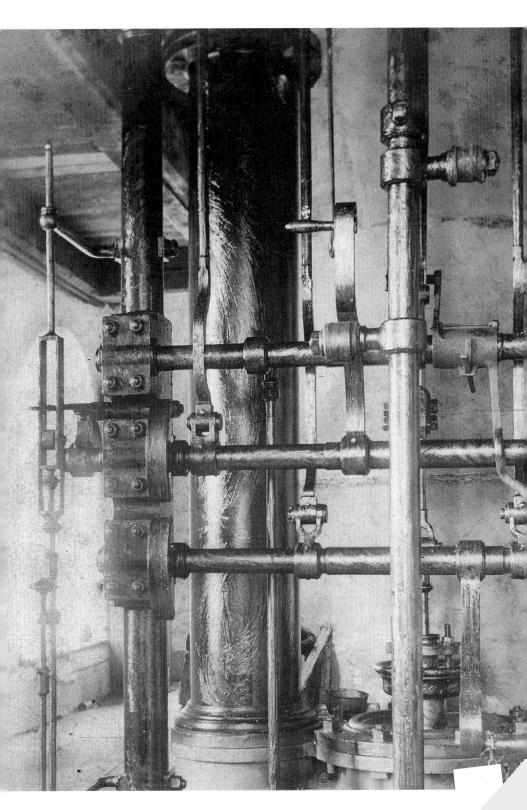

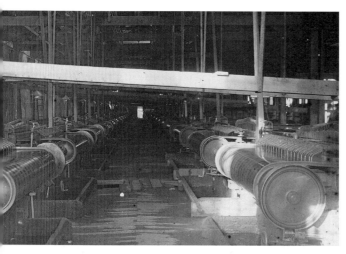

The Frue vanner house at the stamps floors on the western side of the Tuckingmill Valley. The bands of tin concentrate on the vanners are clearly visible on the end roller, showing how rich the ore of Dolcoath was. (*c.* 1910)

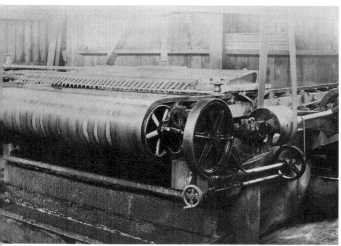

A Frue vanner thought to have been photographed in the dressing floors of the Californian mill on the eastern side of the Tuckingmill Valley. (Late nineteenth century)

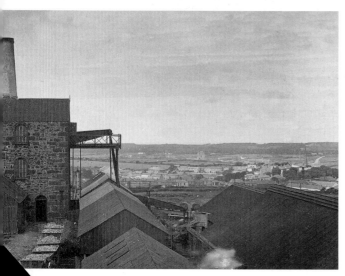

The North stamps engine taken from the bob platform of the South stamps. These are the two which stood on the western side of the Tuckingmill Valley. A man standing by the side tipping tram waggons is about to dump them into the ore bins of the stamps which are under the roofs. The vanner house is on the right. Just beyond the auxiliary pumping beam which returned dressing water can be seen ore waggons on the inclined tramway which connected this old stamps site to the Californian stamps on the eastern side of the valley. (*c.* 1900) (*W.J. Bennetts*)

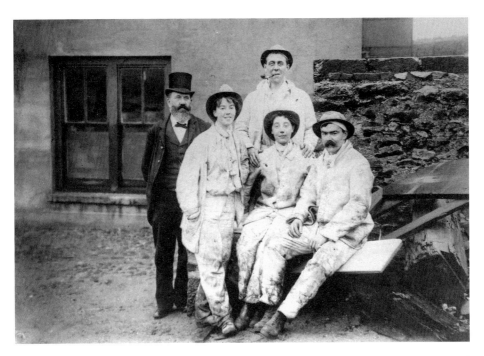

Taken by the counthouse. On the left in formal dress is Captain Josiah Thomas, manager of Dolcoath mine. The man with a pipe is an underground captain who has just returned from taking the three visitors below ground. (*c.* 1900)

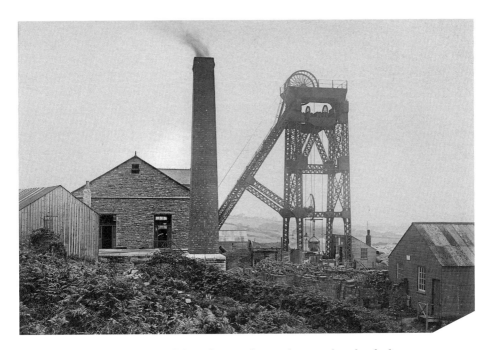

Williams (or New) shaft. Viewed from the east during the time that the shaft was It finally reached a depth of 3,030ft. (1890s)

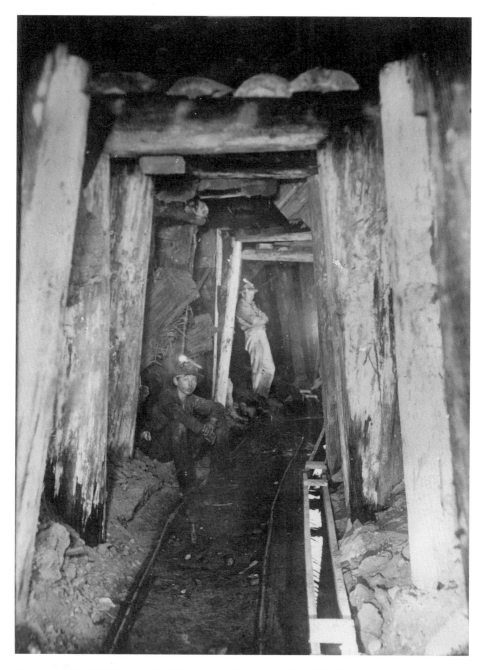

A scene at the 510 fathom level (measured down the dip of the lode, i.e. not vertically). It shows the very heavy timbering required when driving through the Tuckingmill crosscourse. (c. 1900)

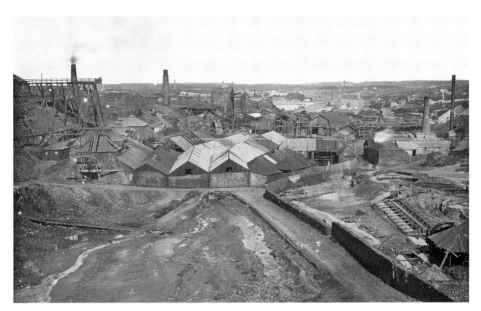

The Tuckingmill Valley, taken from the railway embankment and looking north. On the left is a gantry carrying a tramway for the conveyance of ore from the stamps floors to the calciners, with hip roofs, below. The smoking stack served the calciners. To its right is the stack of the Old stamps, which is seen further to the right. The tapered stack served the boilers of the winding engine used in the sinking of Valley shaft out of the scene. (Late nineteenth century)

Williams shaft. The lattice steel headgear being erected. In the background is a large number of waggons for eventual use underground. (Late 1890s)

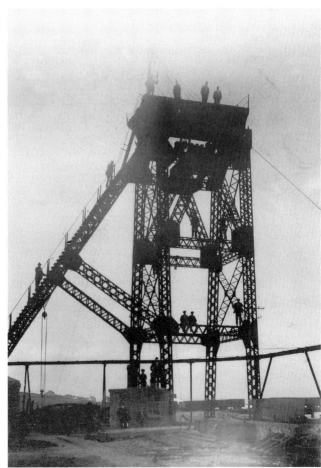

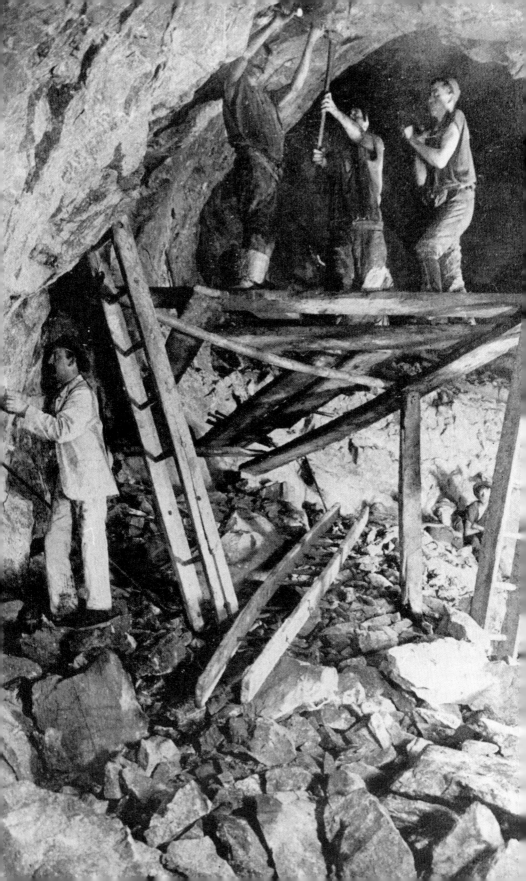

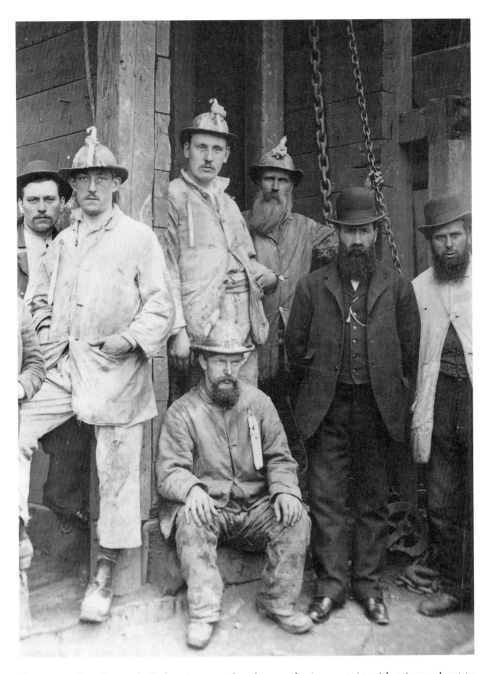

A scene at New Sump shaft showing a surface boss and mine captain with miners about to descend underground. (1893) (*Photograph: J.C. Burrow FRPS*)

Opposite: Taken at the 375 fathom level in March 1904. It illustrates the method of using three ladders as the legs for a stage from which to work. The figure on the left is Captain R.A. Thomas, who later became the principal of the Camborne School of Mines. (*Photograph: J.C. Burrow FRPS*)

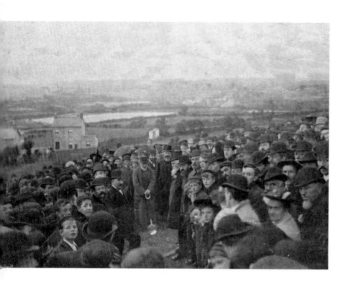

Williams shaft. The cutting of the first sod in 1895. The figure with the ceremonial spade is Mr M.H. Williams, chairman of Dolcoath mine, and the bearded man in a top hat to his left is Captain Josiah Thomas, manager of Dolcoath. *(Photograph: J.C. Burrow FRPS)*

The counthouse yard at the time of the commencement of the sinking of the new Williams shaft. To mark the occasion each man employed by the company received 2 shillings and each boy and girl 1 shilling as a present. At the time (1895) nearly 1,300 persons were employed. *(Photograph: J.C. Burrow FRPS)*

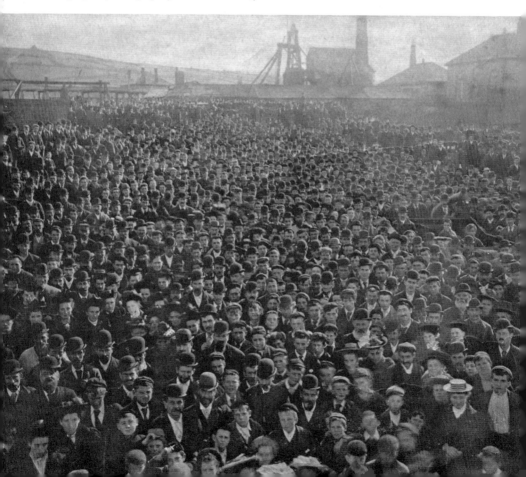

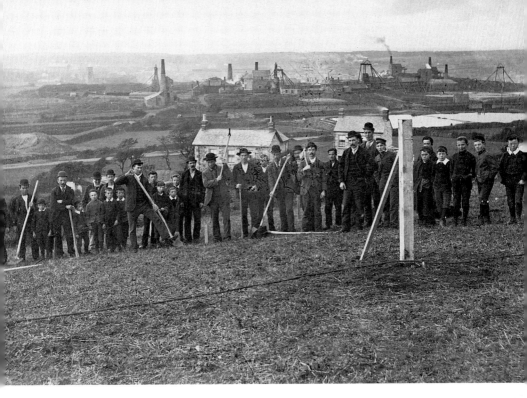

Williams shaft. Taken around the same time as the previous photographs and showing preparations to commence sinking a large rectangular shaft. This was sunk to a depth of approximately 300ft when kaolinised granite was encountered and a decision was taken to convert to a circular brick-lined shaft of 18ft diameter. After much discussion, this was reduced to 17ft 6in and the shaft was sunk to a depth of 3,030ft.

Williams shaft. A scene in the early operations connected with the sinking of this shaft. To the right two men are working on the rectangular setts, which was the original design of the shaft as described previously. It will be noticed that there is a small pumping bob on the left-hand side, presumably operating a bucket-lift pump. (c. 1896)

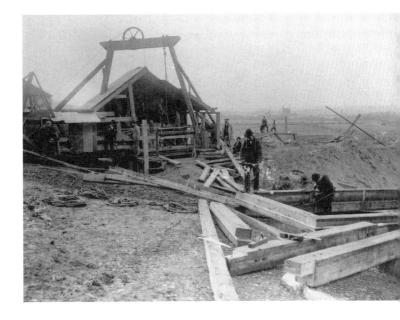

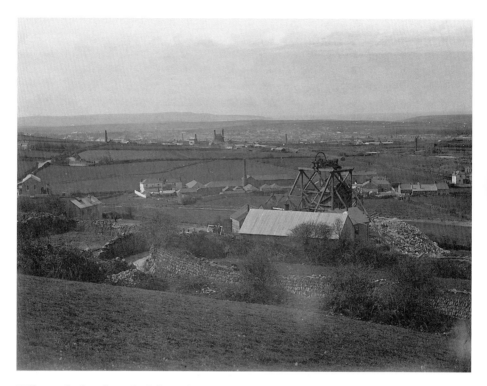

Williams shaft. Taken a little later than the previous print and from the flank of Carn Entral. This shows a more substantial sinking headgear. In the background, left to right, are the stack on Davey's Ropeworks, a headgear on a shaft of Camborne Vean mine, the totally enclosed beam winder, pumping engine and tall shears on Straypark shaft, the stack at Harvey & Co.'s sawmills and finally the iron stack and rotary beam engine which drove the Dolcoath man engine. (*c.* 1897)

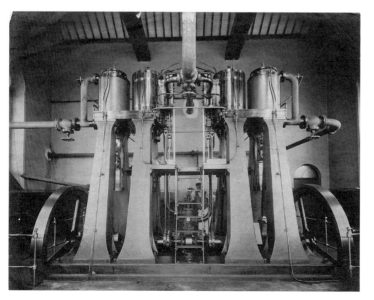

A vertical air compressor which stood near New Sump shaft. This was the first big air compressor to be erected in Cornwall. The building which housed it is still standing. (1890s)

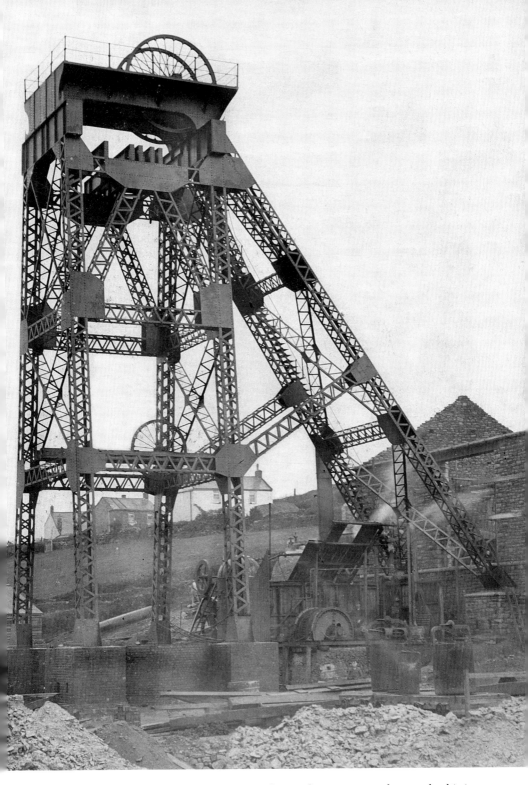

Williams shaft. The permanent headgear is complete and a temporary sheave wheel is in place to serve the small sinking winder. The building to house the traversing winding engine has scaffolding around it. (1890s)

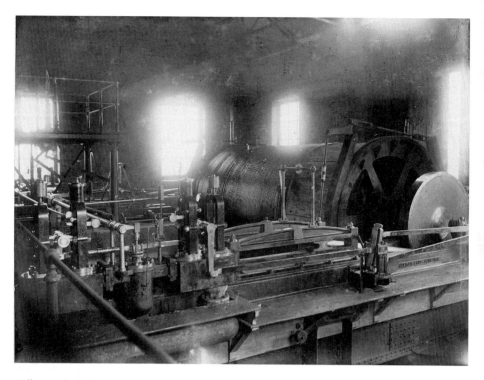

Williams shaft showing an internal view of the traversing winder. This was the only winding engine of its type to be erected in the world. Because the shaft was sited on the side of a steep hill, the winding engine had to be very close to the shaft collar. In order to maintain a true rope alignment with the headgear wheels and correct layering on the winding drum, the whole engine had to move horizontally when operating. This unique machine was designed by T. & W. Morgans, the consulting mechanical engineers to Dolcoath mine. It was said to be capable of hoisting at 3,000ft per minute. The contract for its construction was awarded to Holman Brothers of Camborne. (1912)

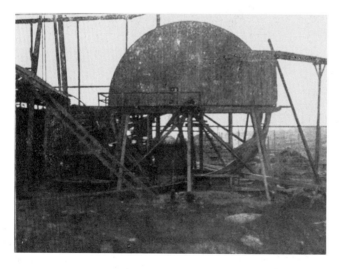

A side view of the large dipper wheel, referred to in a previous scene (p. 14), which raised the pulp from the Old stamps on the western side of the Tuckingmill Valley to the Californian stamps floors on the eastern side. (1890s)

Opposite: A scene at the 326 fathom level showing a rock drill team in the 1890s. (*Photograph: J.C. Burrow FRPS*)

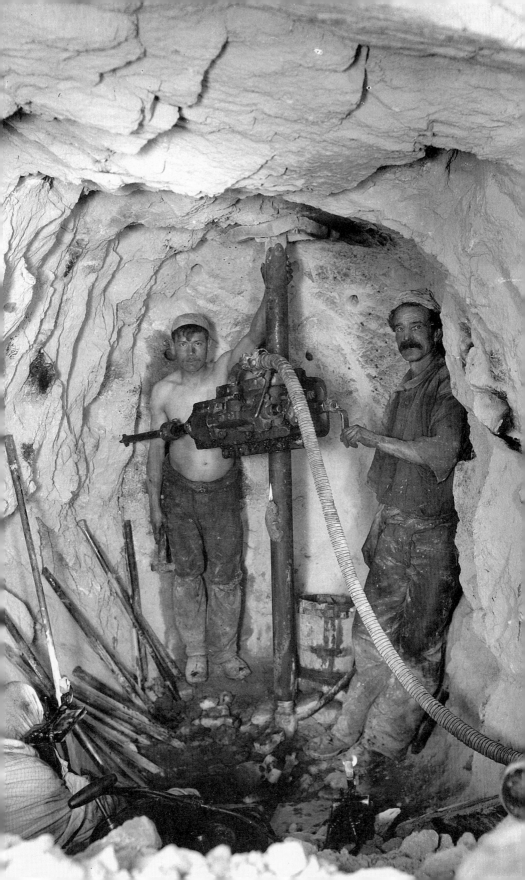

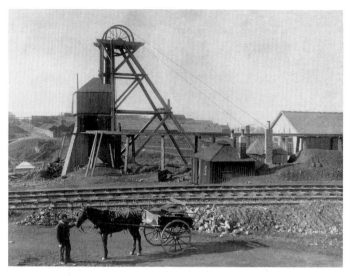

Valley (or Eastern) shaft. The most easterly working shaft in the Dolcoath sett. On the other side of the valley are the Cornish stamps and dressing floors. The railway tracks in the foreground are a part of the GWR branch line to North Crofty mine. (*c.* 1897) *(Photograph: J.C. Burrow FRPS)*

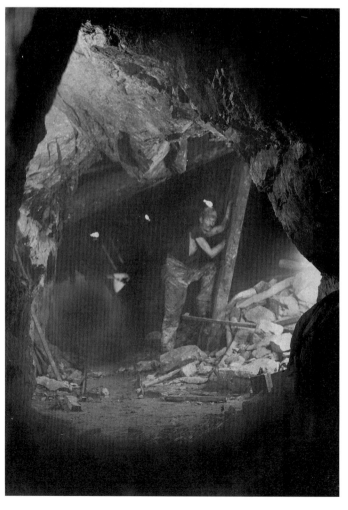

A miner about to climb into a stope at the 388 fathom level. (1890s) *(Photograph: J.C. Burrow FRPS)*

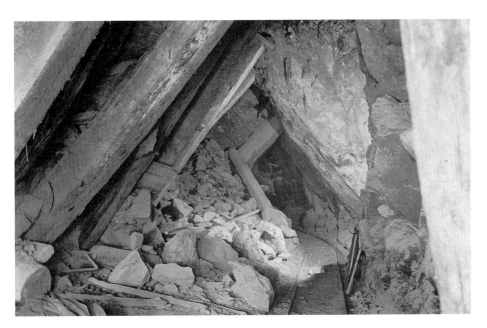

Very heavy timbering of a stull which appears to have broken through. This is at the 388 fathom level. (1890s) *(Photograph: J.C. Burrow FRPS)*

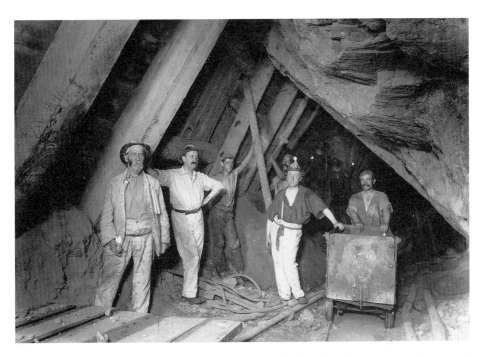

Showing heavy timbering at the 440 fathom level. The bearded figure is an underground captain named Rule. The two central figures are members of the Thomas family, whose father, Captain Josiah Thomas, was at that time the mine manager. (1890s) *(Photograph: J.C. Burrow FRPS)*

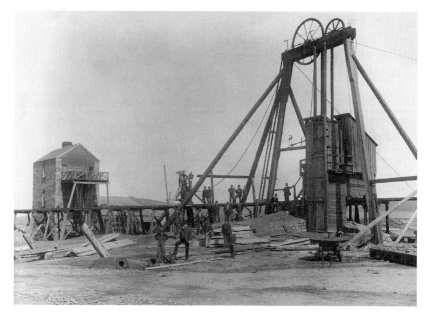

New East shaft. The beam winding engine is on the left with the headgear and landing brace surmounting the shaft whilst the low trestle carried a tramway. The bearded figure in a top hat is the famous Captain Josiah Thomas, the mine manager. (1890s) *(Photograph: J.C. Burrow FRPS)*

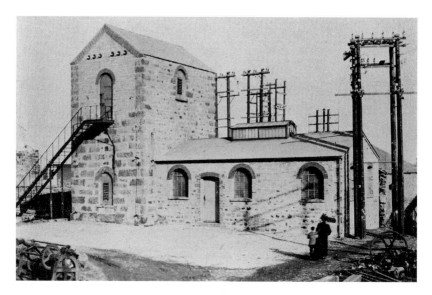

New East beam winding engine house. This shaft had ceased to be used for hoisting and the beam winder, probably one of the oldest pieces of machinery on the mine, was scrapped. Dolcoath was one of the first mines to avail itself of a supply of electricity, which had become available when the Urban Electric Supply Company built their generating station at Carn Brea. The former winder house was converted to an electricity substation. (Early 1900s)

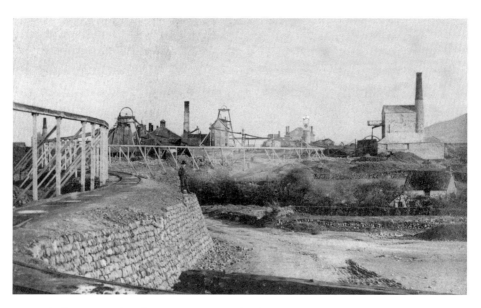

A view looking east taken from a point near Wheal Harriett shaft, one of many shafts served by the tramway. A wooden launder (carrying water) is on the left supported on high posts with struts. In the background, from left to right, are the headgear on Old Sump shaft, New Sump shaft pumping engine stack, engine house and capstan shears, followed by the underground manager's house. Nearer the camera is the mine clock tower and on the right the beam winder which hoisted on Old Sump shaft. (1870s)

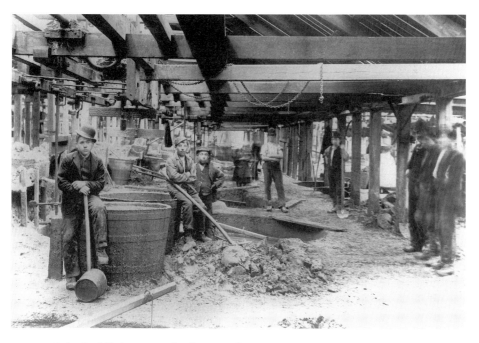

A part of the buddle house in the dressing floors on the western side of the Tuckingmill Valley showing men and boys who are part of this labour-intensive process. (1890s)

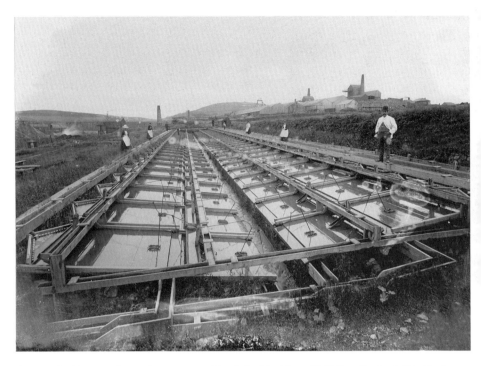

A scene looking south on the western side of the Tuckingmill Valley. On the left is the lower and oldest set of Cornish stamps with its separate stack. The next stack to the right is at the end of the labyrinth from the calciners. Above this are the extensive dressing floors and the two later sets of stamps. The headgear behind the dressing floors is on Gossan shaft. In the foreground are bal maidens and men tending a large set of rag (otherwise rack) frames used for the recovery of fine tin. (*c.* 1890)

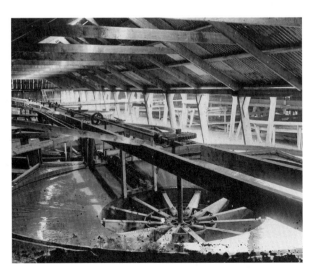

Round frames in the 'fines' section of the stamps floors situated on the western side of the Tuckingmill Valley. (Early 1890s)

*Opposite top:*A part of the dressing floors below the higher stamps on the western side of the Tuckingmill Valley. In the background can be seen the stamps heads whilst in the foreground are buddles. (*c.* 1904)

Opposite bottom: A further view of the old dressing floors at Dolcoath. The buddle by the man in a white jacket is being dug out. Note the bal maidens and young boys among the workforce. It is difficult to comprehend the conditions under which these people worked on very low wages. (Early 1900s)

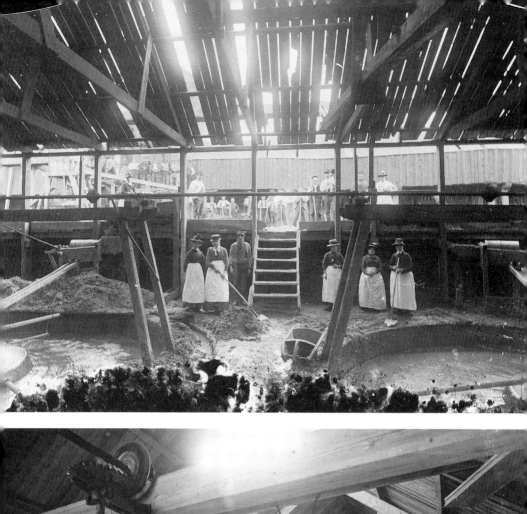

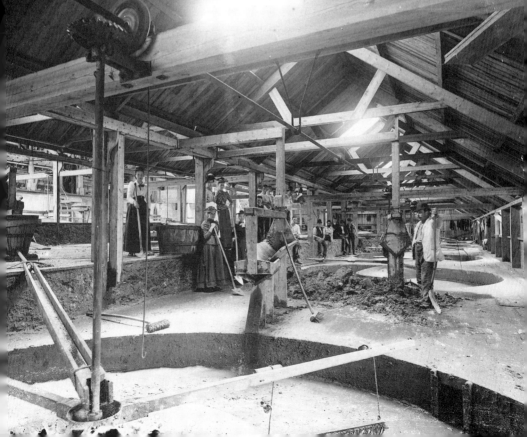

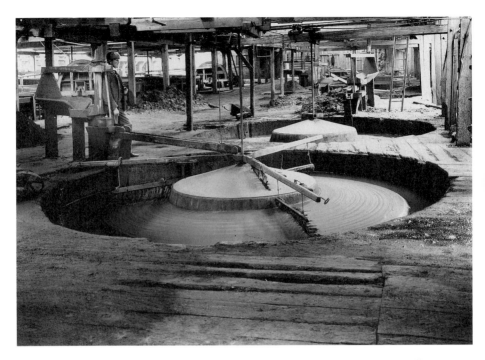

Convex buddles in the dressing plant. The solitary figure in this scene is the mill manager. His name was Thomas but he was not related to the Thomas family long associated with Dolcoath. (*c.* 1904)

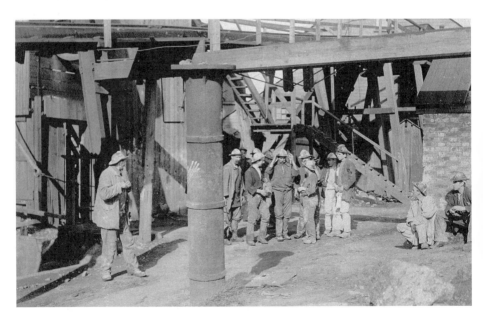

New Sump shaft. A group of miners about to go underground. They are standing under an elevated tramroad. It is amusing to see the handprint on the pump column in the foreground. Clearly someone had been mixing up candle clay and slapped his hand against the pump to wipe off the surplus clay! (Late nineteenth century)

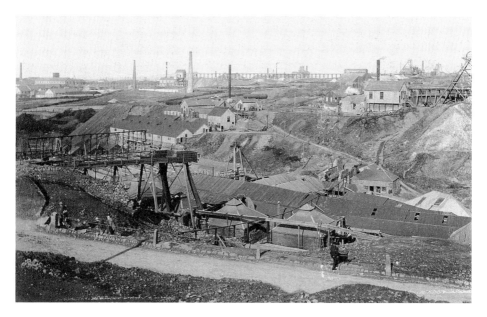

A view across the Tuckingmill Valley looking north. Left to right are the stack and cooling tower of the Urban Electric Supply Company's generating station at Carn Brea, the long building containing the sixty head of Californian stamps at South Crofty mine, followed by two tall arsenic stacks. Between these is the crusher station constructed to serve the New Cookskitchen shaft of South Crofty, then temporarily in suspension. The crushers were utilised to treat ore from Eastern shaft of the old Cookskitchen mine served by the long gantry leading to the shaft headgear on the skyline. In the middle ground left is the Dolcoath Californian stamps and to the right Valley shaft winding engine and headgear. In the foreground are the Dolcoath calciners and part of the dressing floors. (*c.* 1910)

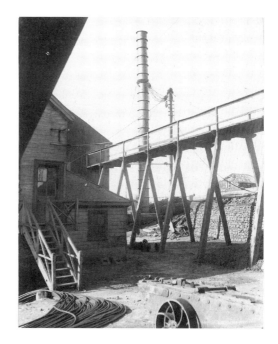

Valley shaft. This view, taken from under the headgear, shows the steps leading into the horizontal steam winder house. A spare shaft skip is in the foreground and a replacement steel stack has just been erected. The tramway trestle linked the shaft landing brace to the crusher station, which is just visible to the right of the winder house. (Early twentieth century)

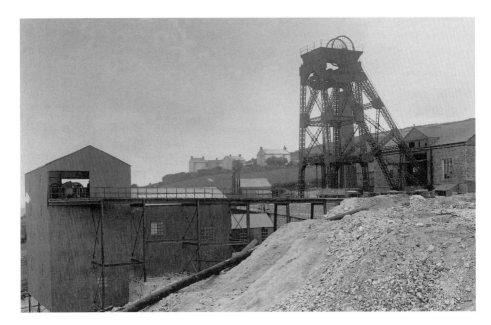

Williams shaft. This photograph was taken when the plant was fully commissioned and showing on the left the waggon tippler plant with one of the gantries linking it to the shaft. The terminal poles for the electricity supply are in the centre middle ground of the print. Behind these is a building that contained the electric winch, which operated the inclined tramway. (c. 1913)

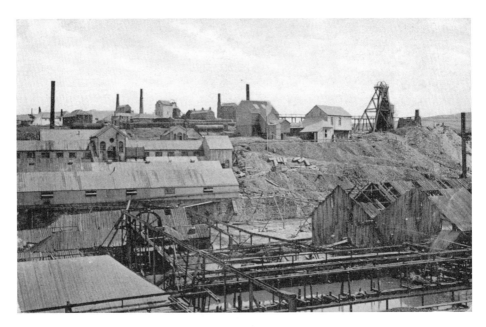

The eastern side of the Tuckingmill Valley. On the left is the Californian stamps and to the right the crusher station, winding engine house and headgear at Valley shaft. In the background is Cookskitchen mine. (c. 1900)

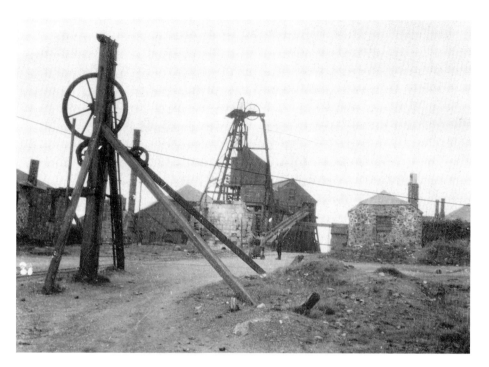

New Sump shaft. Taken in 1922 after the mine had closed. The steel headgear and capstan shears are still in place as is the incline to the crusher station on the right. In the foreground is a pulley stand carrying the rope from the Old Sump shaft beam winder to the headgear. Much of the surface equipment of the mine remained in place for a very long time and not until the early years of the Second World War was the final clearance completed.

Old Sump shaft in 1922. This shows the headgear and winding engine with the trestle from the landing brace to the ore bin, which is out of shot on the right. The mine had ceased work by this time. Note how the skip is blocking the mouth of the shaft as a safety precaution.

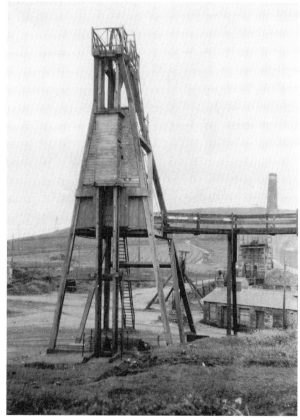

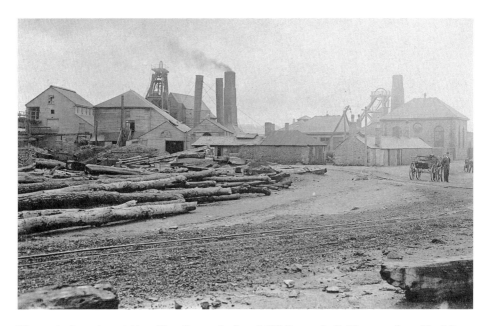

The central crusher station, New Sump shaft and Old Sump shaft. The complex of buildings also includes the main workshops of the mine and the blacksmith's shop. The tall building on the right contained the first large air compressor to be installed in Cornwall. Part of the extensive 1ft 10in gauge steam-operated tramway can be seen in the foreground. (*c.* 1912)

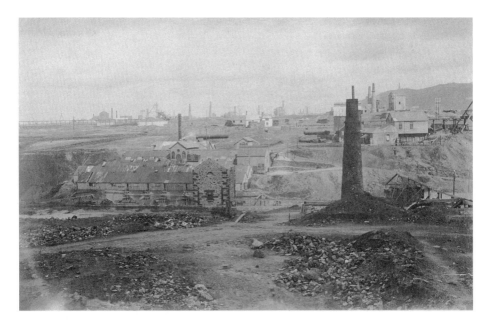

The caption on this postcard sums up several of the mines in this view. The Californian stamps of Dolcoath mine are on the left. Part of the Valley shaft complex is on the right and in the background are Cookskitchen mine, Tincroft mine, South Crofty mine and the Carn Brea mines. (*c.* 1915)

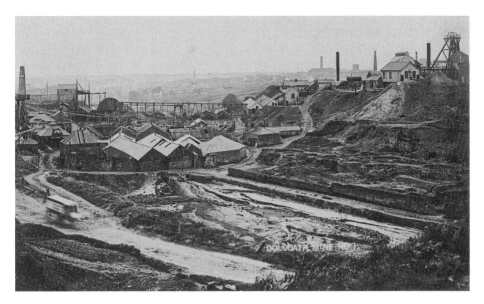

A view taken from the western side of the Tuckingmill Valley looking north. Old stamps and the very large dipper wheel are on the left. The Valley shaft complex is on the left of the photograph and the Californian stamps is situated in the centre of the scene. (*c.* 1910)

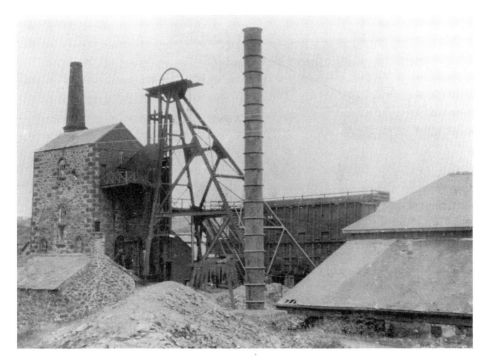

Wheal Harriett shaft. The house contained a 65in Cornish pumping engine. At this time a new steel headgear had been erected together with new ore bins. A new horizontal steam winder completed the updating of the plant on this shaft. (*c.* 1897)

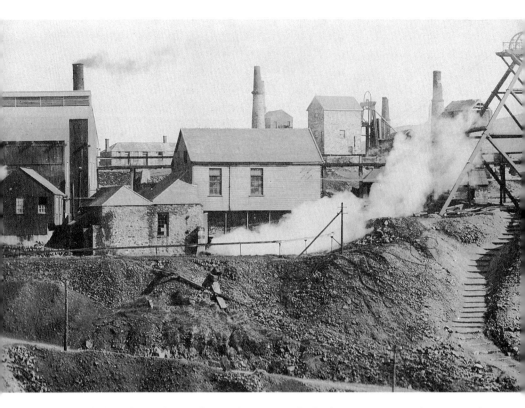

Valley shaft. Left to right are the crusher station, boiler house, steam winding engine house and headgear. The stack for the boilers, which served the winding and crusher engines, is smoking. The miners' steps, which were a feature of the valley, can be seen. There was a similar set of steps on the western side of the valley. In the background is the complex of buildings around Chapple's shaft of Cookskitchen mine. (*c.* 1900)

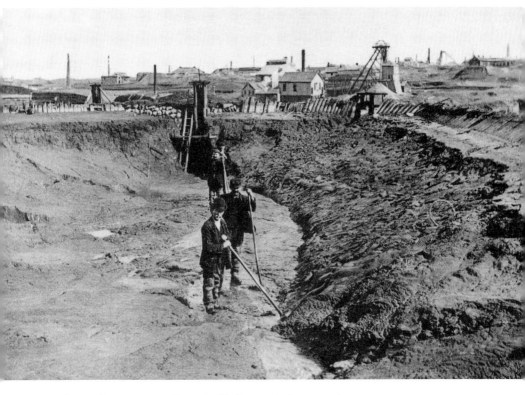

A team of men cleaning out a slime pit. (Early twentieth century)

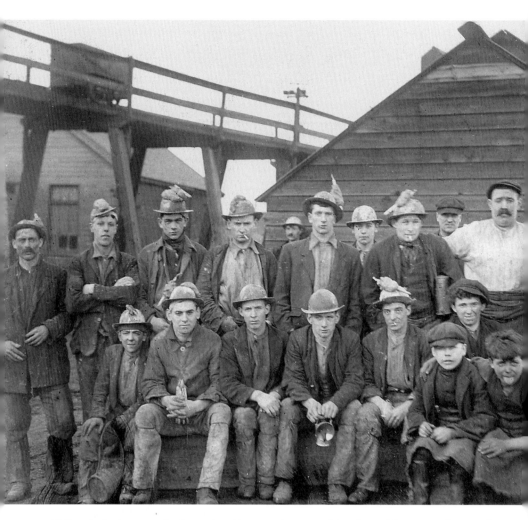

Valley shaft. A group of miners about to go underground together with some surface workers and boys. All have candles except the boss who is holding an acetylene lamp. The man second from the left in the front row is Frederick Uren. He emigrated to the USA prior to the First World War and was killed underground in Painsdale, Michigan, in the 1920s. (*c.* 1911)

Opposite: Straypark shaft. Earlier in this chapter mention was made of a replacement pumping engine for this shaft. This illustration from a Holman Brothers Engineers catalogue shows the middle chamber of the new engine house and engine. (1900)

CORNISH PUMPING ENGINE.
SECOND FLOOR.

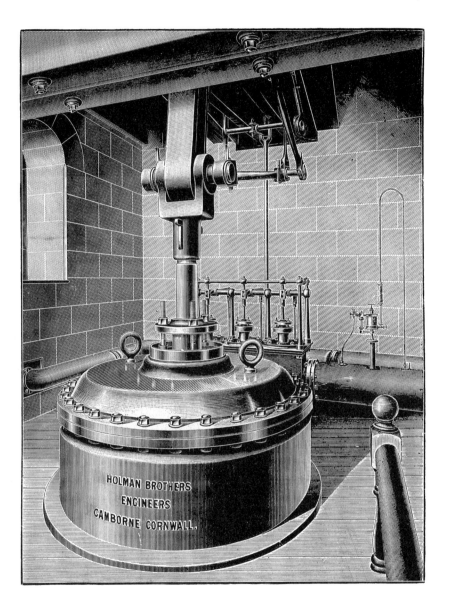

The above engraving shows the Top nozzle, Piston rod, and Parallel motion connections.

ESTIMATES ON APPLICATION.

CORNISH PUMPING ENGINE.

THIRD FLOOR.

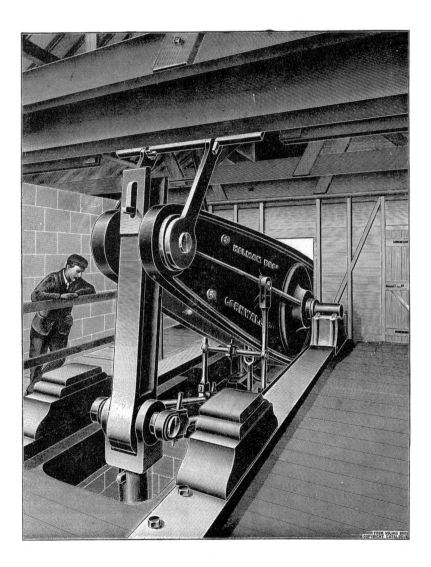

The above illustration shows the main connection to Beam, &c.

ESTIMATES ON APPLICATION.

Straypark shaft. A further page from the Holman catalogue showing the bob loft. The new engine had a 65in cylinder. It was decided to retain the bob of the old engine and consequently the illustrator's lettering on the beam is technically incorrect. (1900)

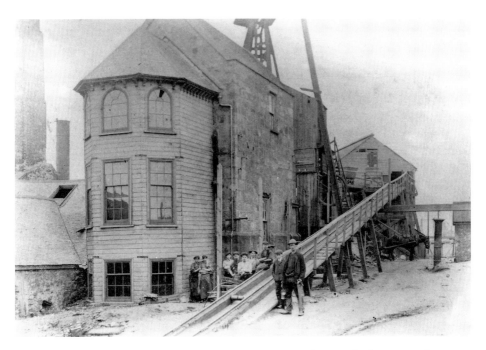

New Sump shaft showing the boiler house and the engine house which contained an 85in pumping engine together with the headgear and capstan shears. The inclined tramroad leads to the central crusher station. (Early twentieth century)

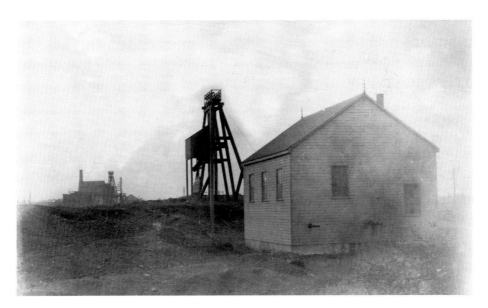

Wheal Killas shaft showing the headgear and winding engine. This shaft was re-equipped in the early twentieth century with a view to mining the shallow low-grade copper and zinc ores to be treated at the Elmore Oil Flotation Plant mentioned earlier in this chapter. The venture was not a success. In the background is Straypark shaft, the most westerly working shaft on the mine.

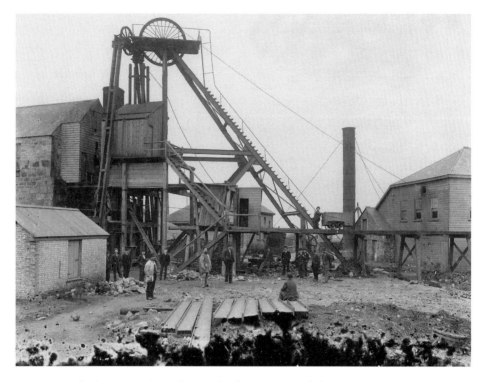

New Sump shaft. A view from the south showing, from left to right, the 85in pumping engine house, the recently erected new steel capstan shears, steel headgear, winding engine house and in the background the compressor house. (Late 1890s)

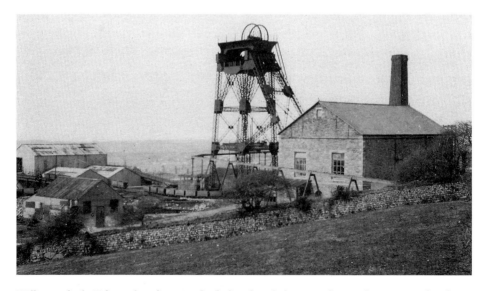

Williams shaft. Taken after the mine had closed and showing the crusher station, headgear and traversing winding engine house. The stack served the boilers which steamed both the winder and air compressor. There are a large number of ore waggons which had been brought up from underground when operations ceased. (1922)

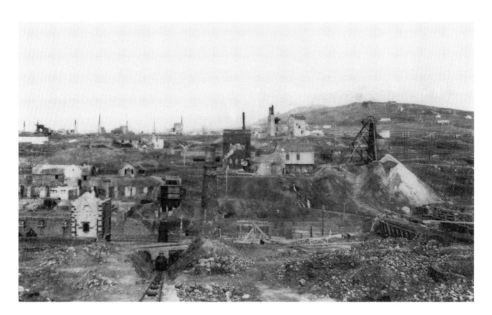

A view across the Tuckingmill Valley looking east after the mine had stopped working. In fact, all the mines in this scene had closed by this time. The inclined tramway was installed when the Californian stamps, seen derelict on the left, had ceased working as an economy measure. The ore from Valley shaft, seen on the right, was then conveyed to the stamps on the western side of the valley via this incline. (*c.* 1922)

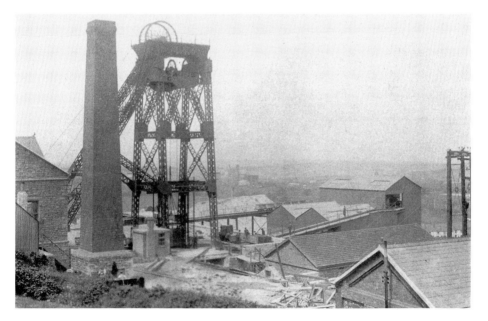

Williams shaft. A view of the winding engine house, stack, headgear and crusher station. The boiler house is to the right of the stack and on the edge of the print are the electricity terminal poles which brought power to the site and in particular for the electric pumps. The building in the right-hand corner contained the electric hoist on the tramway incline. (*c.* 1912)

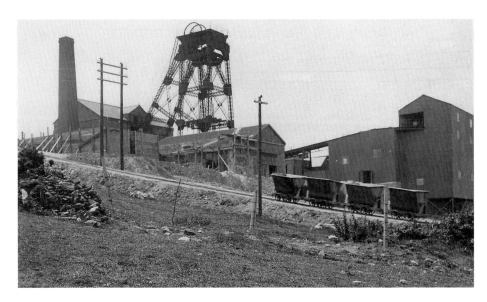

Williams shaft. Taken from the north-west and showing waggons on the incline. An additional Lancashire boiler was being added to the original two. The scaffolding for the boiler house extension is in evidence. To the left of this is the shuttering for the foundations of the large Corliss valve air compressor being built by Holman Brothers of Camborne. Both the winder and compressor worked at a steam pressure of 125lb psi. (*c.* 1912)

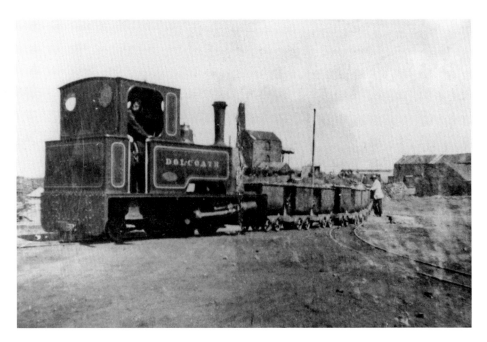

This shows the first of the two Kerr Stuart locomotives which worked the Dolcoath mine tramways from 1903 to 1920. They were 22in gauge 0-4-2 tank engines. The tracks on the right lead to the central crusher station situated near New Sump shaft. (*c.* 1906)

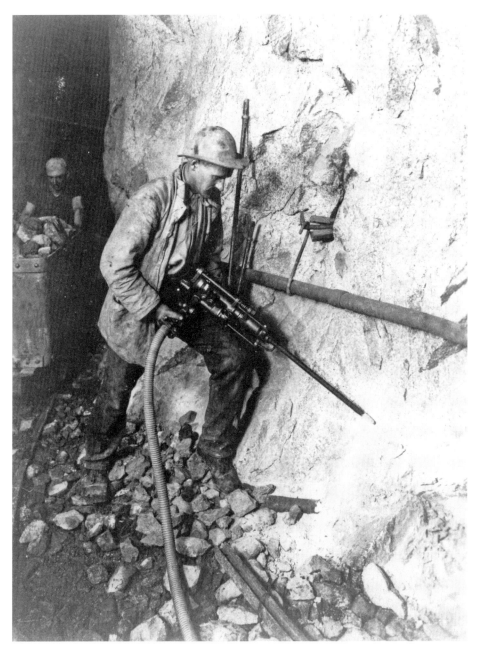

A hand-held pneumatic drill in use at the 440 fathom level near Williams shaft. (*c.* 1913)
(Photograph: J.C. Burrow FRPS)

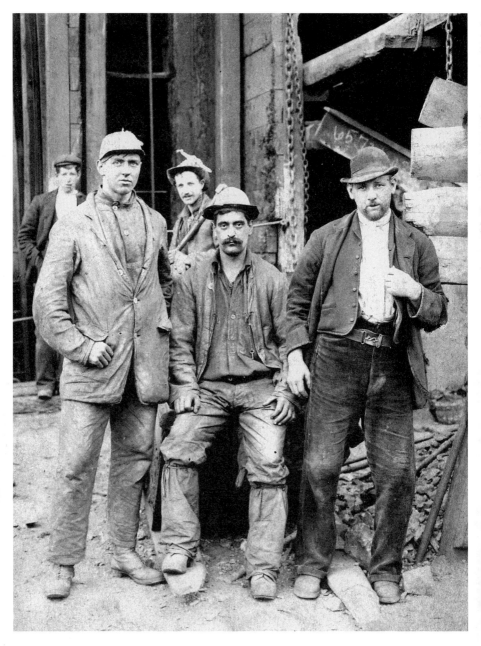

New Sump shaft. A group of miners showing the winding compartment of the shaft immediately behind them. The seated figure is James Dunn. (*c.* 1910)

Opposite: Somewhere in France during the First World War. This shows four Royal Army Medical Corps NCOs, all of whom were former Dolcoath mine employees. They had been members of the mine ambulance team prior to the outbreak of war. The centre rear figure is Dick Olds. The captain is Dr Blackwood, who had a medical practice in Camborne in pre-war times. (*c.* 1915)

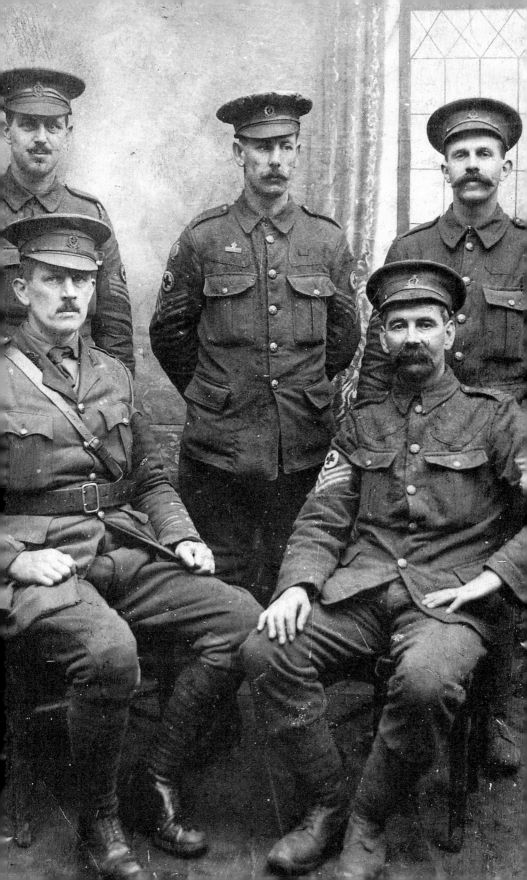

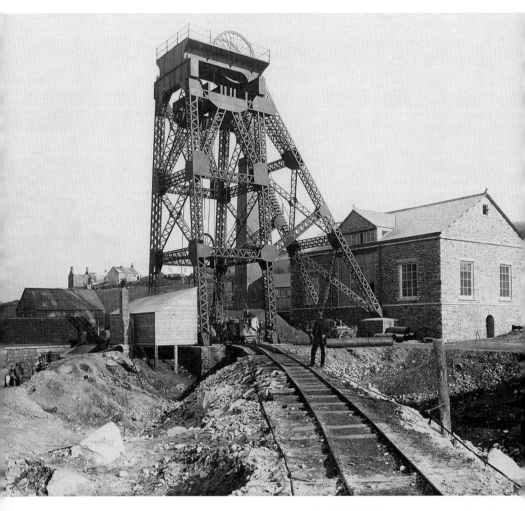

Williams shaft. At the time the shaft was being sunk. The lofty lattice permanent headgear has replaced the sinking headgear and the traversing winding engine is now being used for the sinking operations. This unique winder actually commenced working in July 1900. (1908)

Opposite: Wheal Harriett compressor house. This stood at the eastern end of Dolcoath Avenue, which is the terrace appearing in the background. The engine was built by Holman Brothers in 1904. (1907)

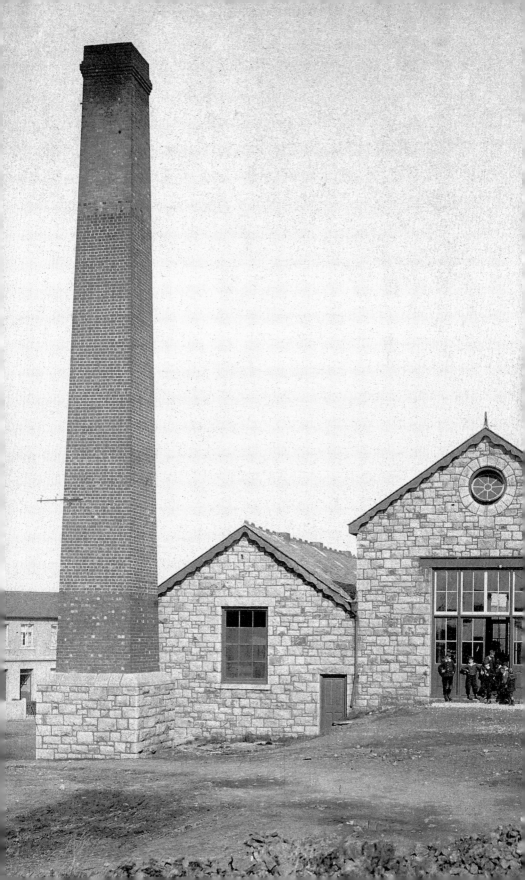

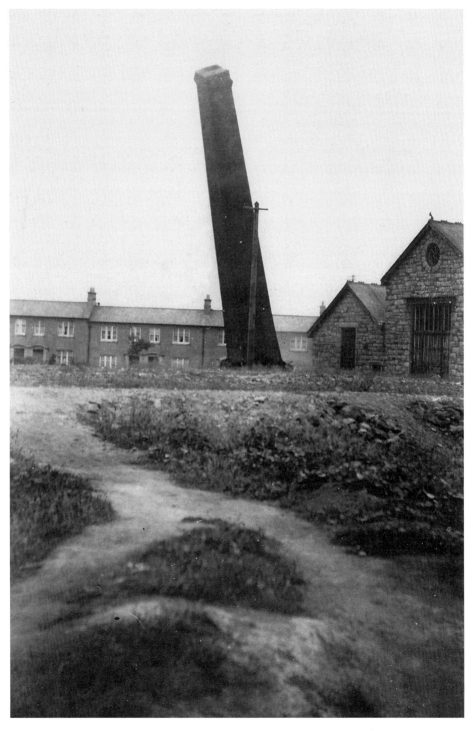

A much later view of the compressor house seen in the previous photograph taken at the time the brick stack was felled. The recovered bricks were used to build the stack at New Dolcoath mine at Roskear which is still standing. (Early 1920s)

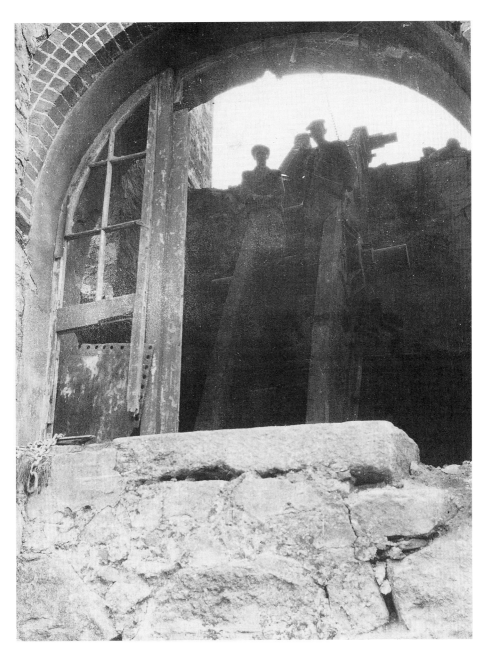

Wheal Harriett shaft 65in pumping engine when being scrapped. The bob has been blasted off the wall and will be broken up inside the house. This engine had originally worked at the Wheal Harriett mine near the village of Beacon. It was purchased by the Dolcoath Company and re-erected on this shaft where it worked for many years until the closure of Dolcoath. (*c.* 1940)

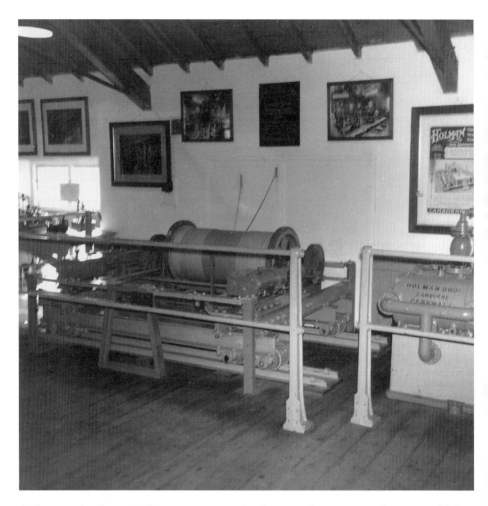

A photograph taken in Holman Museum at Camborne in the 1970s. It shows a model that demonstrates the principle of the traversing winding engine which the company built for the Williams shaft at Dolcoath mine in the late nineteenth century. *(T.A. Morrison)*

Opposite: Williams shaft. The Ministry of Supply required a water supply for the establishment being erected at the former RAF airfield at Portreath. The late J.H. Trounson was consulted and he suggested that the specifications and quantities of water might be obtainable from this shaft. The print shows the tower erected over the shaft for access purposes and the pipes carrying the water from the submersible pumps to the measuring weirs during the experimental stage. It transpired that this source satisfied the needs of the Ministry and a pipeline was laid from the shaft to Portreath, and remained in use for many years. (Late 1940s)

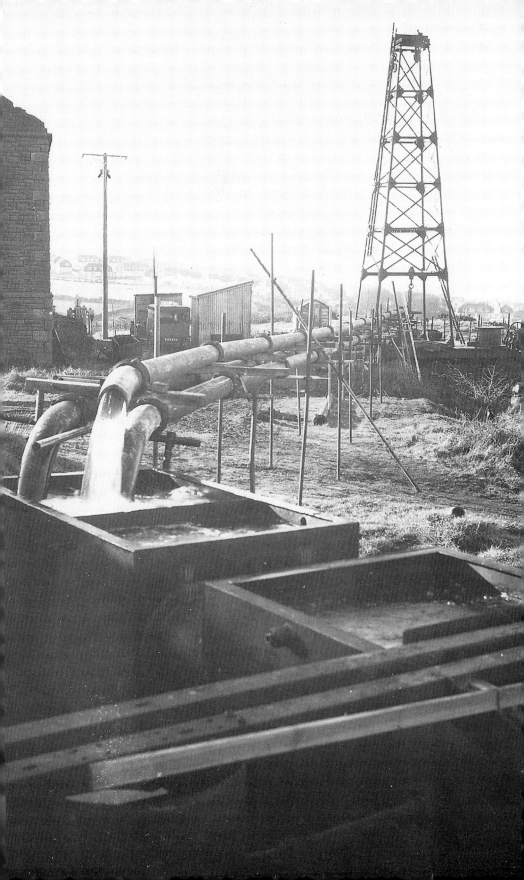

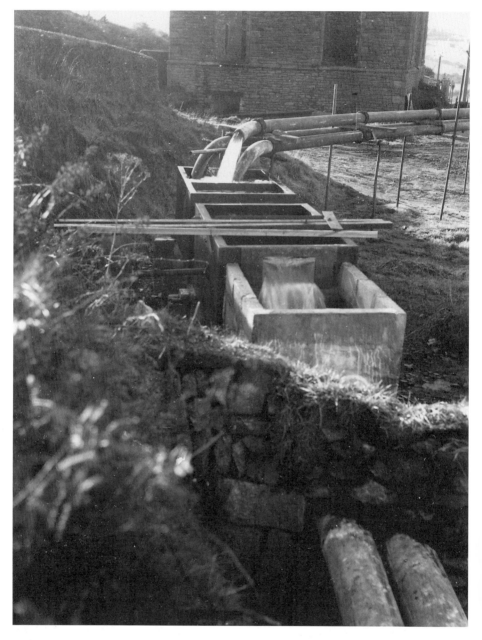

Williams shaft. A further view of the plant described in the previous photograph and showing the building which had contained the traversing winding engine. (Late 1940s)

2

New Dolcoath Mine

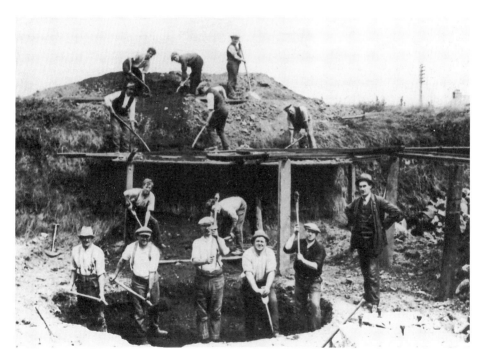

The commencement of the sinking of the new shaft which was 16ft in diameter within the brickwork lining. This was originally intended to be sunk to a depth of 1,680ft but in fact was completed to 2,000ft. A firm of mining contractors from Wales secured the contract for the shaft sinking. (1923)

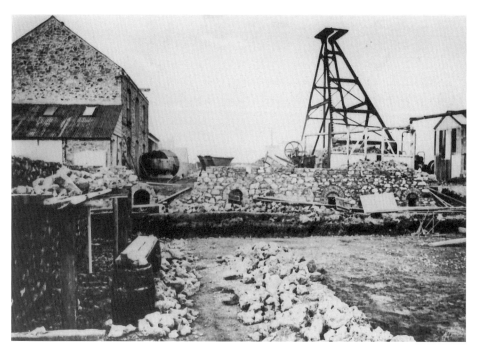

A view from the south showing the sinking headgear which had been brought from the Dolcoath mine and re-erected here. (1923)

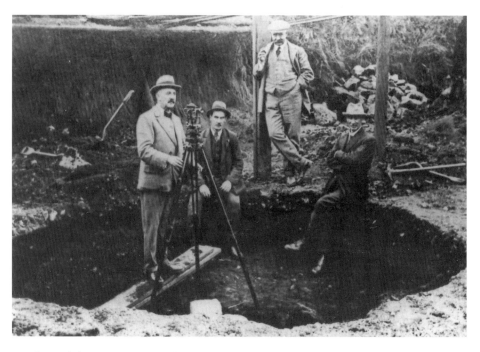

Members of the contractors' staff, one of whom is surveying, at the shaft collar which had been excavated to a depth of about 3ft. (1923)

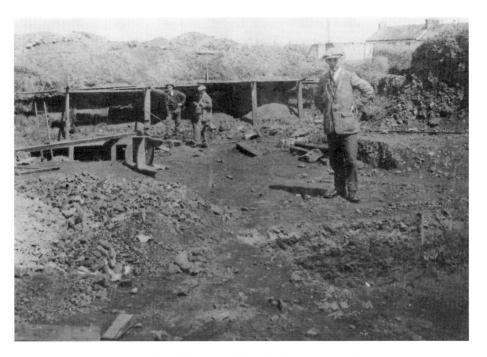

Site preparations near the collar of the new shaft. The figure nearest the camera is Joe Pearce, the site clerk, who was a Dolcoath mine employee. (1923)

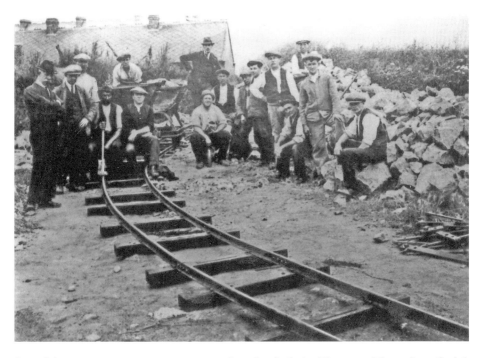

One of the temporary tramways constructed on the shaft site. The second figure from the left is Joe Pearce. (1923)

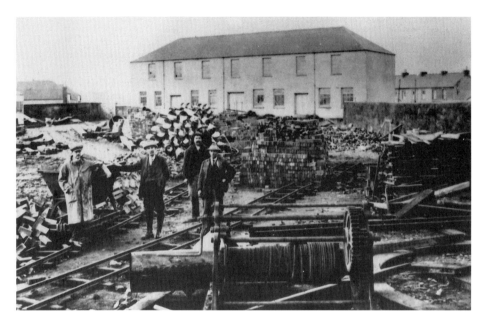

Showing the storage of building materials and equipment using the temporary tramways. The pile of bricks in the centre of the photograph had almost certainly been salvaged from the compressor stack referred to in chapter one. (1923)

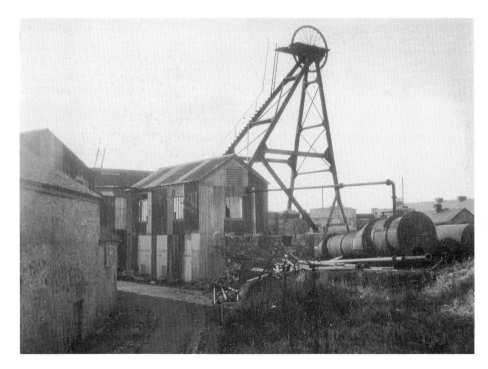

This is the temporary sinking headgear. The galvanised iron-clad shed contains a small Holman air compressor and in the shed behind is the horizontal steam winder. A Cochran boiler was used to steam both of these engines. (1923)

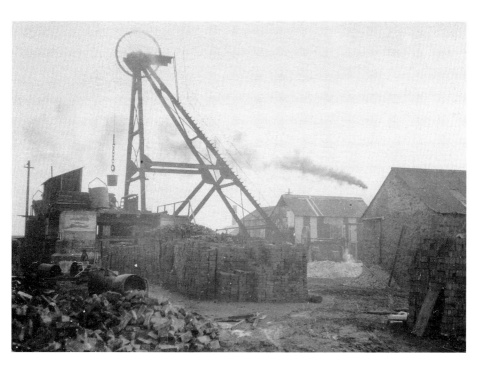

A view looking east showing the diminutive sinking bucket used in the initial stages of shaft sinking which was the limit of the capacity of the small sinking winder. It will be noted that a much larger bucket is awaiting the installation of the permanent winder. The shaft collar has been formed about 15ft above ground level in order to give adequate dumping room for the rock from the shaft sinking in the field on the other side of the road. Eventually the whole of the vast dump that accumulated was bought by the county council and crushed on site for use as road stone and the field was returned to agricultural use. Note the large piles of new bricks which were to be used for lining the new shaft. The Cochran boiler is smoking copiously. (1923)

An early view of the site showing the erection of the various temporary plant and buildings. A small air receiver is on the left, with a pipe leading to the compressor house, and the temporary headgear is being built. (February 1923) *(Photograph courtesy of G.A. Jones)*

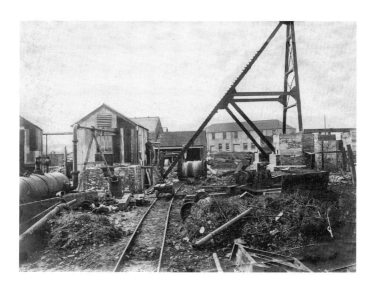

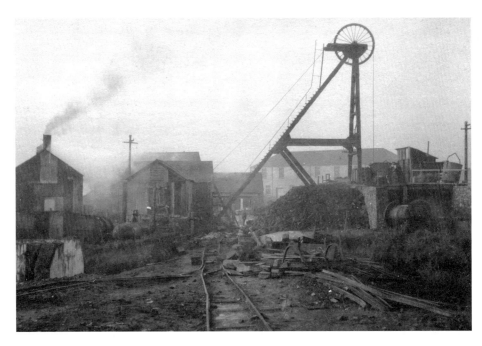

A scene taken looking west. Left to right: boiler house, compressor house, winder house and headgear. A certain amount of material has come up from the shaft and on the right is one of the two geared drums of the sinking capstan awaiting erection. In the background is the large stores erected many years previously by the Bennetts Fuse Works. There is an interesting note with this print that reads: 'taken on Camborne Feast Day, November 1923'.

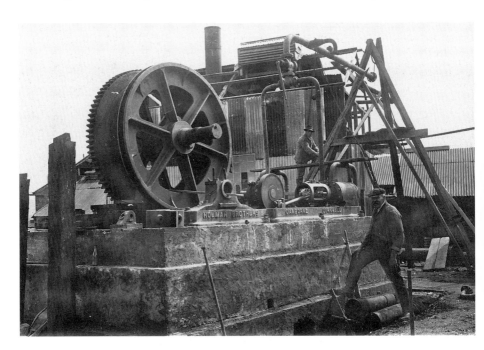

The sinking winder being erected. (August 1923) *(Photograph courtesy of G.A. Jones)*

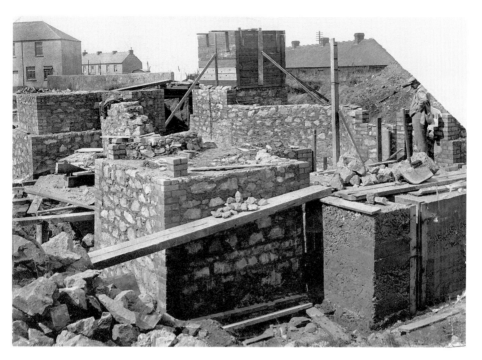

Foundations for the permanent headgear being built. (September 1923) *(Photograph courtesy of G.A. Jones)*

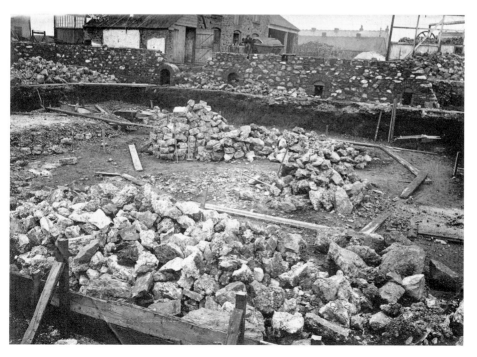

Construction of the boiler flues viewed from the south. (September 1923) *(Photograph courtesy of G.A. Jones)*

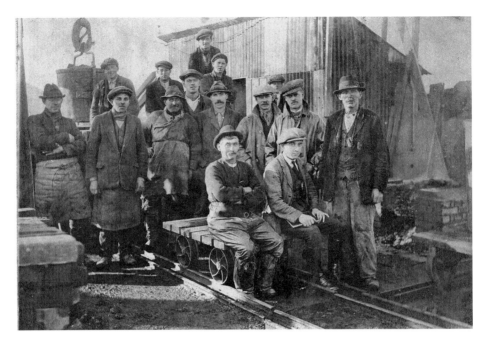

A group of workers by the shaft collar. The man wearing a cap is Joe Pearce, the site clerk. Note the sinking bucket in the background. (December 1923) *(Photograph courtesy of G.A. Jones)*

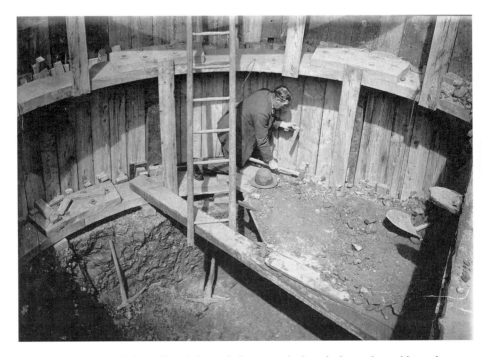

Showing a close-up of the collar of the 16ft-diameter shaft with the timber tubbing that was used down to the level at which the foreman is about to drive in a wedge. Below this it can be seen that the shaft was entering harder ground. (1923)

This team of men is 'shammelling', i.e. raising the excavated material by stages. The print is a good illustration of the size of a 16ft-diameter shaft. (1923)

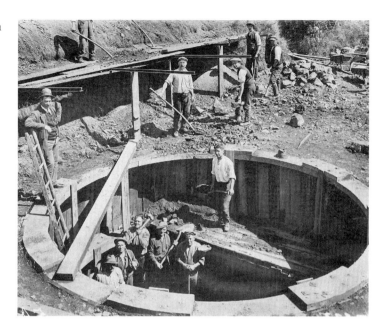

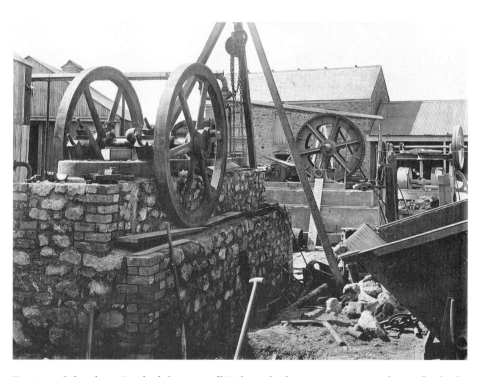

Erection of the plant. On the left is a small Holman-built air compressor with two flywheels. Under the tripod the sinking winder is being erected. In the right background is the steam-driven pug mill for grinding up the ashes and the lime to make black mortar which will be used for all the buildings and the brick lining of the shaft. (1923)

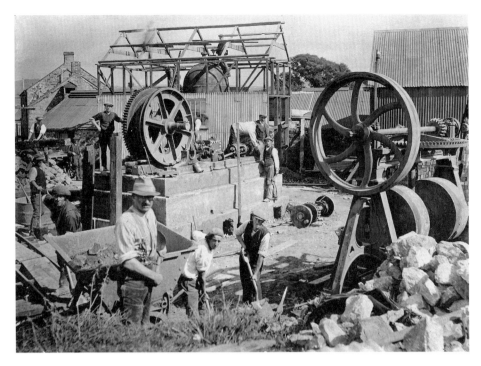

A different view of the operations described in the last photograph. In the background is the top of the Cochran boiler with the shed being built around it. (1923)

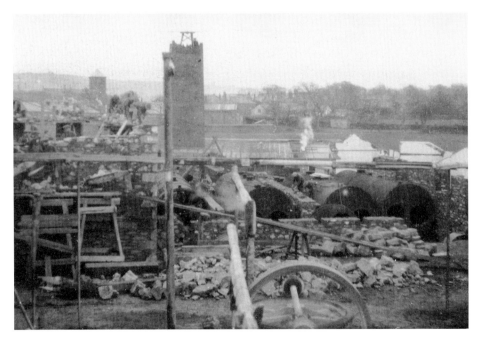

Showing the installation of the four Lancashire boilers and the construction of the stack. (1924)

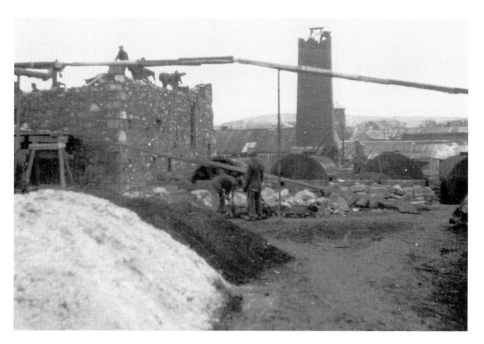

The house for the permanent winding engine is being built and the four Lancashire boilers are in place. (1924)

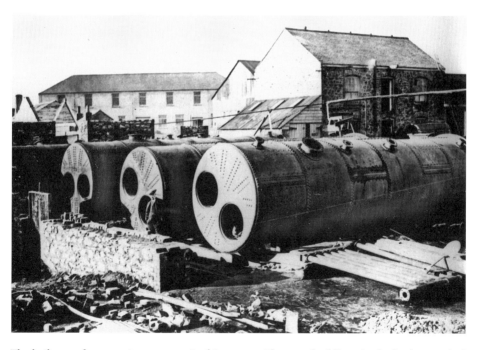

The boilers and economisers appear in this scene with many buildings in the background of the former extensive Bennetts fuse factory. (1924)

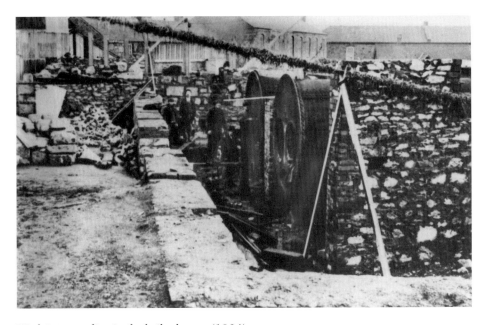

Work is proceeding in the boiler house. (1924)

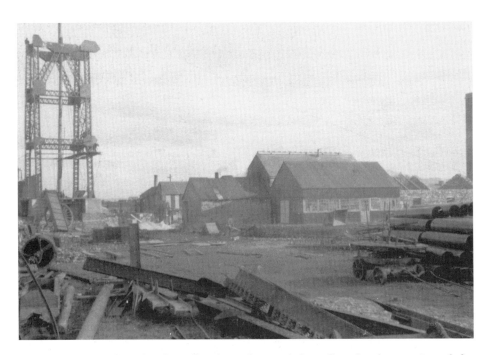

The temporary sinking headgear has been dismantled to allow for the erection of the permanent headgear. This lattice-work structure had been originally erected new at the Williams shaft of Dolcoath mine in the late 1890s. In this photograph, taken from the west, is a partially glazed building and a number of other buildings, all of which were a part of the fuse factory that formerly occupied this site. The new boiler house is being built on the right of the photograph and the stack has been completed. (1924)

A view from the south-east illustrating the sheer size of this headgear as can be judged by the steel erectors seen within the framework. (1924)

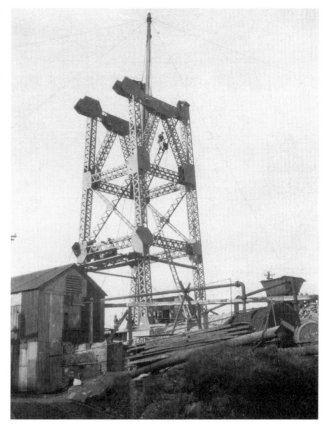

The roof is being laid on the permanent winder house. Two of the four Lancashire boilers are in place. (1924) *(Photograph courtesy of G.A. Jones)*

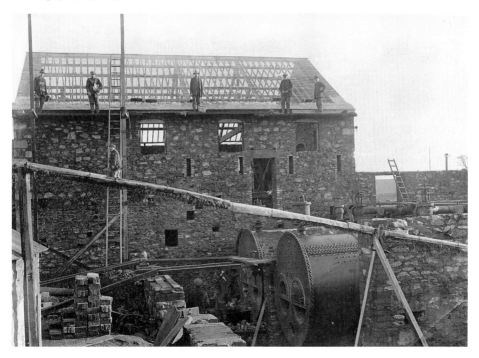

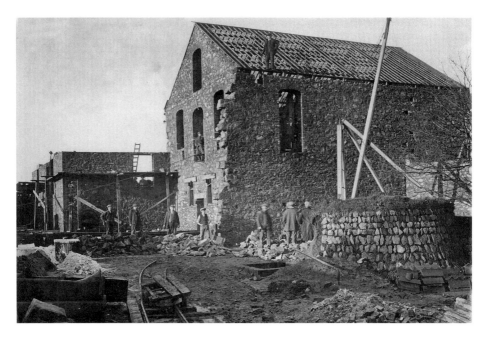

Taken from the east. In the background the boiler house is completed to the eaves and the permanent winder house has the roof timbers in place. (1924) *(Photograph courtesy of G.A. Jones)*

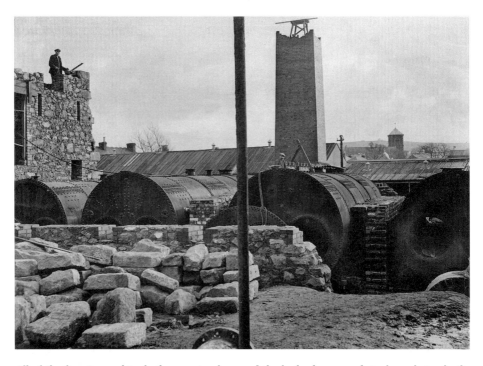

All of the four Lancashire boilers are in place and the boiler house and stack are being built. (1924) *(Photograph courtesy of G.A. Jones)*

Showing the stack under construction. The economisers for the boilers are on the right whilst on the left are buildings of the former fuse factory. (1924) *(Photograph courtesy of G.A. Jones)*

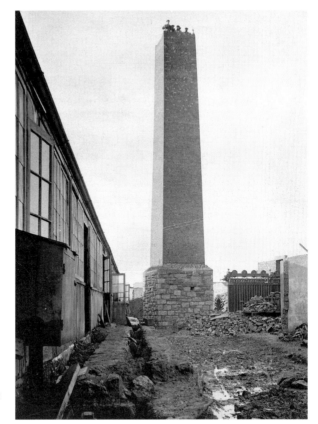

The stack is complete. The building on the left and those being demolished on the right are part of the Bennetts fuse works which was previously sited here. (1924) *(Photograph courtesy of G.A. Jones)*

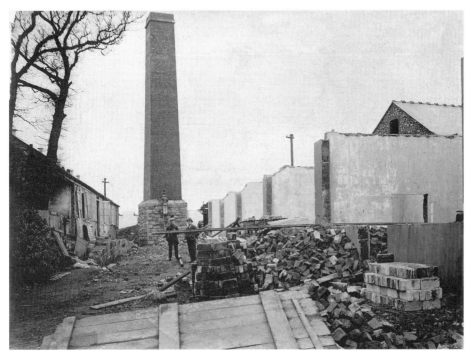

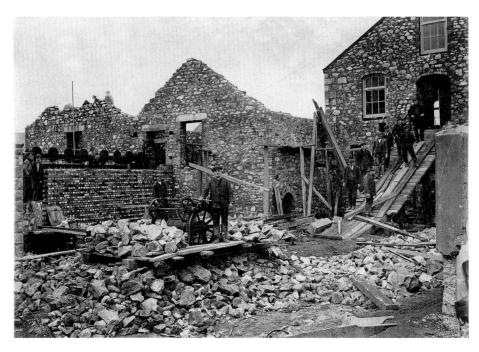

The economisers, boiler house and winding engine house. (14 February 1924) *(Photograph courtesy of G.A. Jones)*

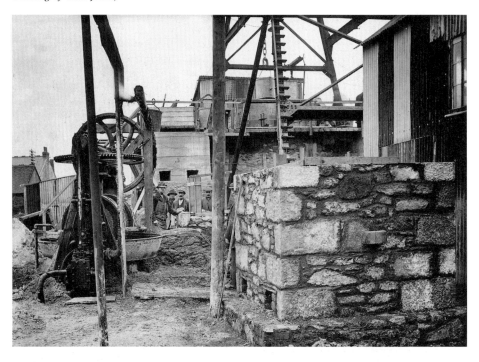

On the left is the steam-driven pug mill described previously and on the right is the foundation for one of the boom stay legs of the permanent headgear. It will be noted that the mortar jointing is black. (1924) *(Photograph courtesy of G.A. Jones)*

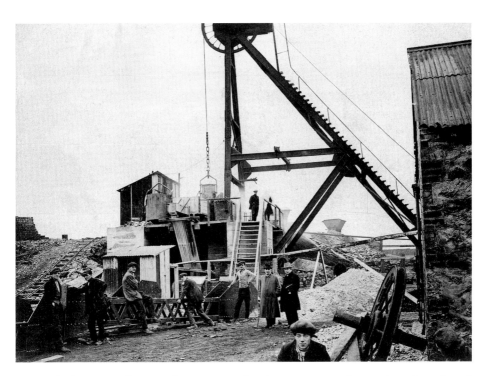

A view looking east showing the temporary headgear in use with a section of one leg of the permanent headgear awaiting erection. The two well-dressed gentlemen are directors of Piggott's, the shaft-sinking contractors. (1924) *(Photograph courtesy of G.A. Jones)*

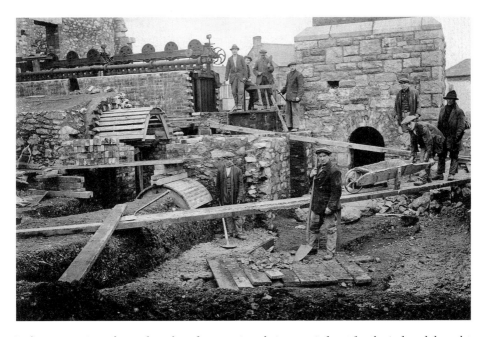

Boiler economisers, base of stack and excavations being carried out for the induced draught fan. (1924) *(Photograph courtesy of G.A. Jones)*

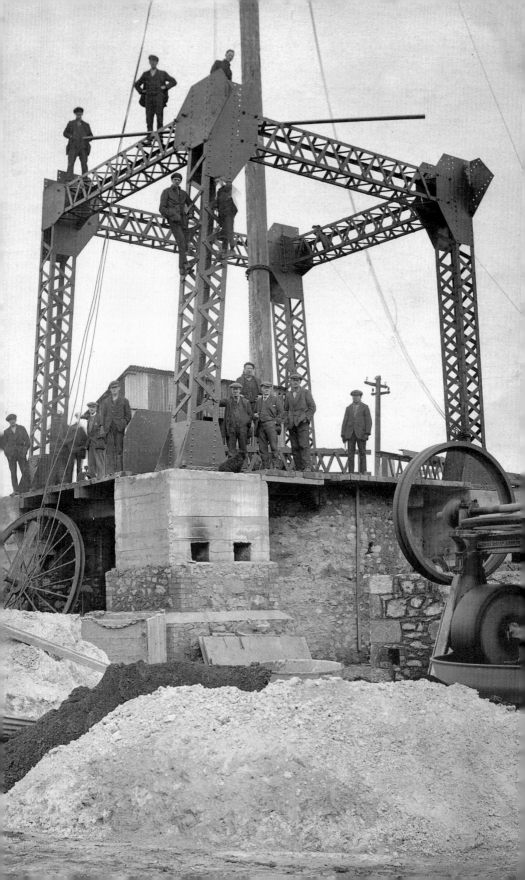

Opposite: The first section of the permanent headgear has been erected. The wheel being raised on the left will be placed at a low level in the headgear as a temporary measure so that shaft sinking can continue using the sinking winder. On the right the pug mill is at work. (1924) *(Photograph courtesy of G.A. Jones)*

Right: The two left-hand legs of the permanent headgear are at their full height and one of the cap pieces, which support the sheave wheel platform, is on the ground. *(Photograph courtesy of G.A. Jones)*

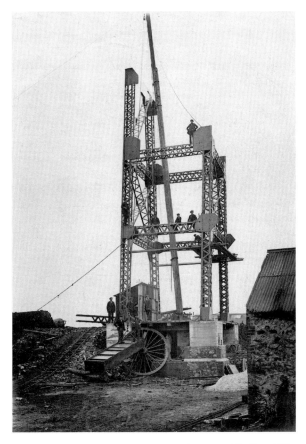

Below: The southern end of the boiler house and winding engine house. (6 March 1924) *(Photograph courtesy of G.A. Jones)*

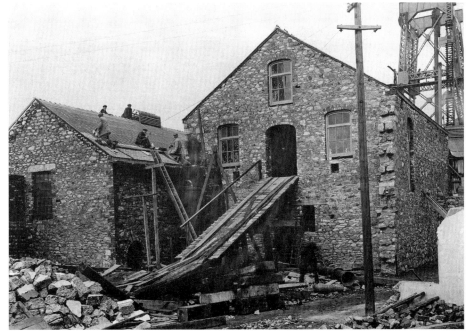

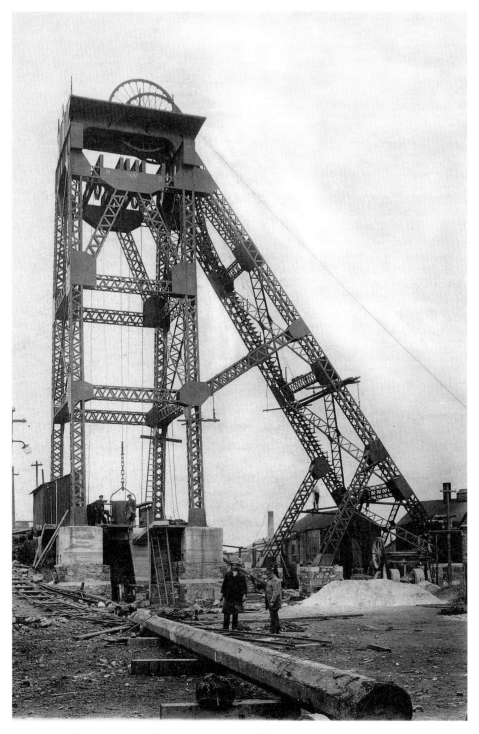

It is early March and the permanent headgear is now in use, together with the permanent winder for the shaft sinking. Note the stack in the background which served the now closed fuse factory. (1924) *(Photograph courtesy of G.A. Jones)*

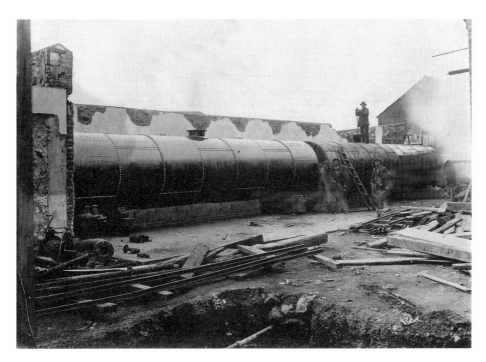

Compressed air receivers being installed. (1924) *(Photograph courtesy of G.A. Jones)*

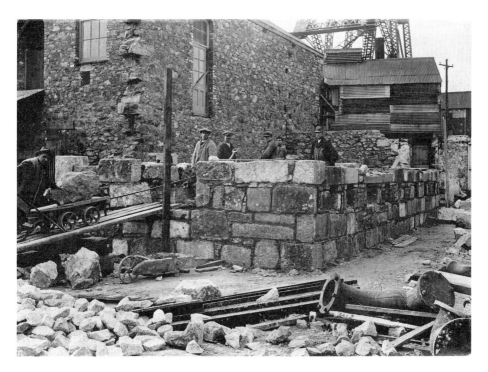

The compressor house is being built. The large steam compressor from the Williams shaft of Dolcoath mine was eventually re-erected here. (1924) *(Photograph courtesy of G.A. Jones)*

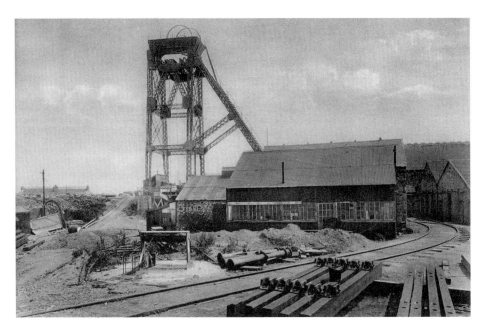

A general view from the west in which the original central building has been adapted into a carpenter's shop. The railway tracks are a part of a short spur laid in from the North Roskear branch of the GWR, thus enabling coal to be delivered straight into the boiler house. However, the curves were so sharp that a locomotive could not be used so the trucks had to be hauled the short distance by horses. (1924)

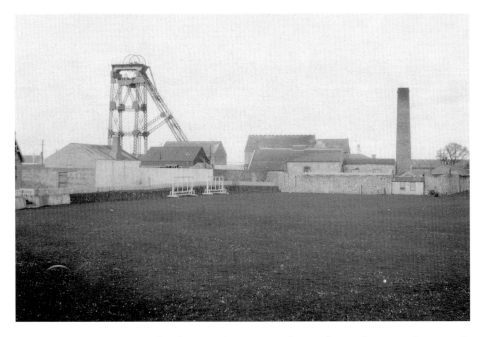

Taken from the south-west. The photographer was standing in the Camborne cricket ground. (1924)

The mine had lain idle for many years when, in 1941, the headgear was felled in the drive for all scrap metal in the Second World War. This shows the main structure at the moment of falling.

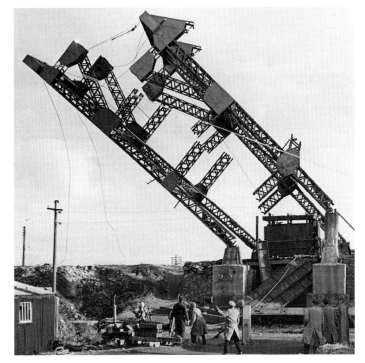

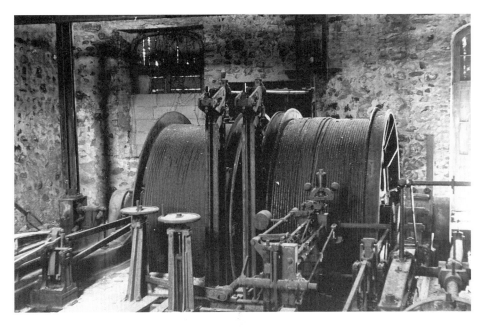

The large steam winding engine which had been bought new in 1897 for the Wheal Harriett shaft of Dolcoath mine and worked there for twenty-three years. It was re-erected here in 1924, when a second winding drum was added together with other work, which included the fitting of post brakes and overwind gear. This photograph was taken in 1974 when the engine had lain idle for forty-five years. *(Photograph: Ron Earl)*

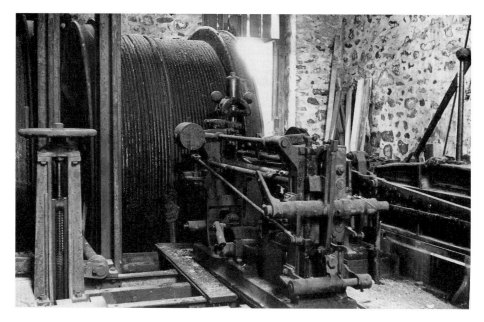

A further view of the winding engine showing the overwind gear and the clutch handle and linkage to the left-hand drum. (1974) *(Photograph: Ron Earl)*

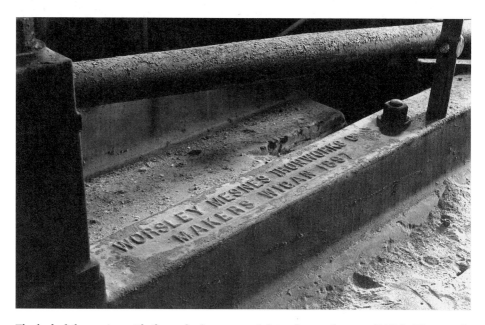

The bed of the engine with the maker's name and date of manufacture. (1974) *(Photograph: Ron Earl)*

3

Other Local Mines and Companies

Cookskitchen Mine

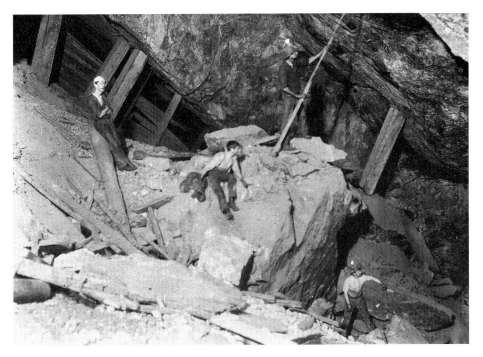

Taken at the 368 fathom level. A timberman is using a slide staff to measure for a new length of timber. (1893) *(Photograph: J.C. Burrow FRPS)*

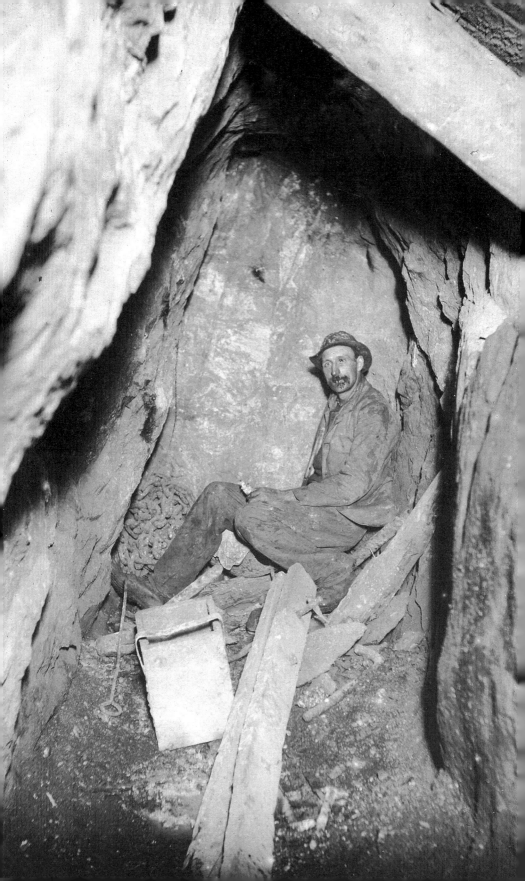

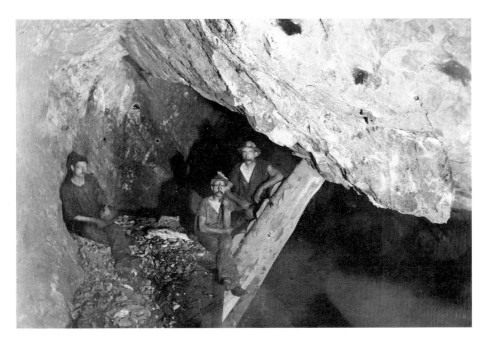

Three miners sitting on a stull at the 355 fathom level. (1890s) *(Photograph: J.C. Burrow FRPS)*

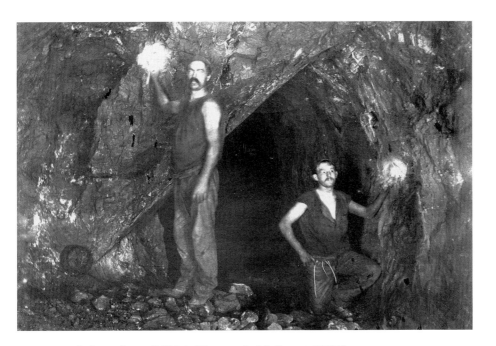

Two miners lighting fuses. (1890s) *(Photograph: J.C. Burrow FRPS)*

Opposite: An end at the 234 fathom level showing a miner charging up his holes before blasting. (1890s) *(Photograph: J.C. Burrow FRPS)*

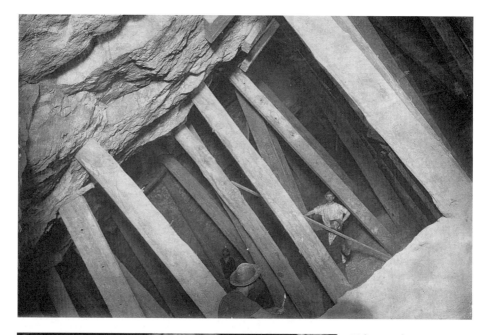

Taken at the
406 fathom level
showing heavy
timbering. (1890s)
*(Photograph: J.C. Burrow
FRPS)*

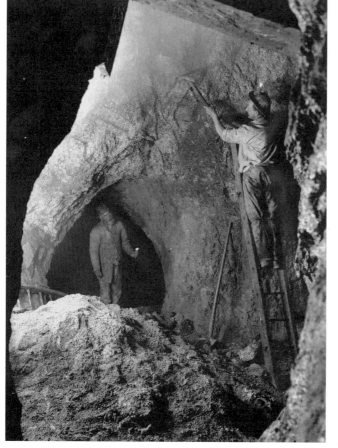

A shift boss has just
arrived on the scene at
the same mine where
a miner is at work at
the 355 fathom level.
(1890s) *(Photograph:
J.C. Burrow FRPS)*

Right: Overhand stoping at the 355 fathom level. (1890s) *(Photograph: J.C. Burrow FRPS)*

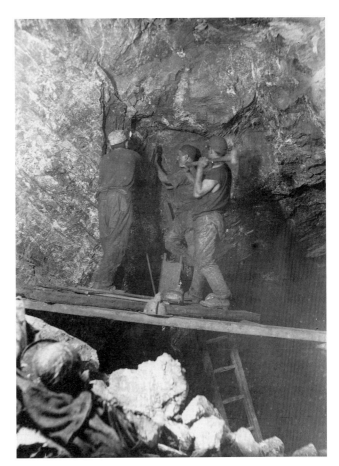

Below: This scene is taken from the western side of the village of Brea and shows, from left to right, Chapple's shaft 55in pumping engine house, the headgear and beam winding engine. This was the principal shaft on Cookskitchen mine. From this point the remainder of the scene is made up of the stacks, engine houses and headgears etc. of North and South Tincroft mines. In the right-hand background can dimly be seen further structures on the Carn Brea mine. (Late nineteenth cenutry)

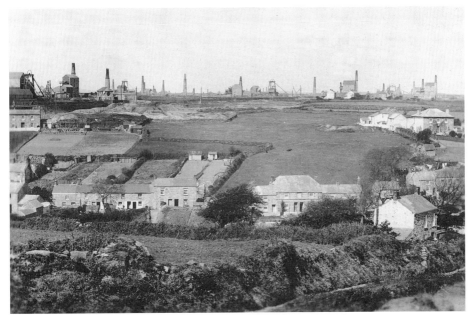

Tincroft Mine

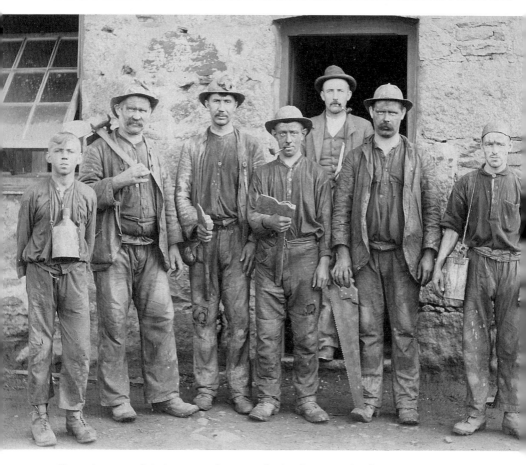

Above: A group of timbermen taken outside the dry. Note the dags (axes), saws and water bottles. The man standing in the doorway is the dry attendant. He stokes the fires which provide the heat to dry the clothes of the miners and generally keeps the place tidy. (1890s)

Opposite top: Downright shaft. A group at the shaft collar. Back row: (unknown), Capt John Hammill, (unknown), Capt George Nancarrow, Hartley Teague, Joe Teague, (unknown), (unknown), Capt W. Teague junior (Billy), George Hosking. Front row: (unknown), Robert Hampton (lander), (unknown), (unknown), John Richards, Phil Sedgeman, (unknown), Charlie Jose, (unknown), (unknown). The term 'lander' should be explained: he was in charge of the cage or skip at the shaft collar and in the landing brace within the headgear. He was also responsible for all the signals to the winder driver from the surface. (Late 1890s)

Opposite bottom: Downright shaft. A photograph taken at the same time as the previous scene showing more of the shaft collar arrangements. (Late 1890s)

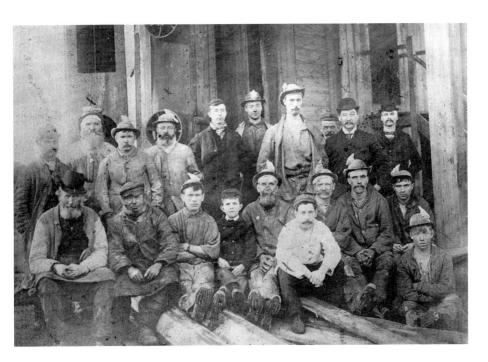

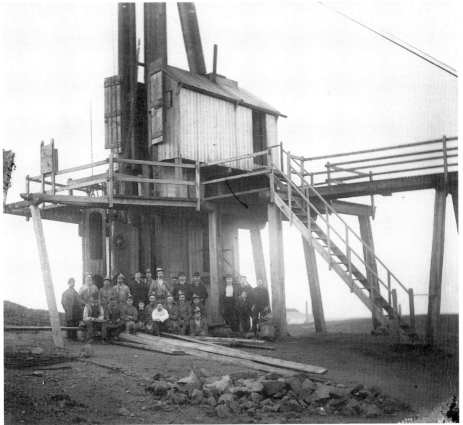

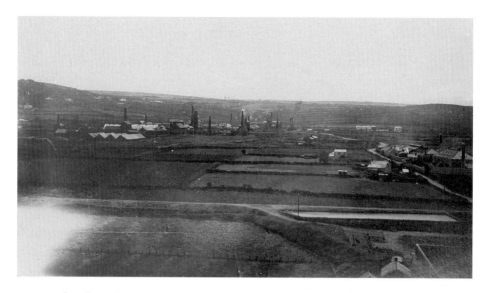

A view taken from the top of the then newly built stack of the electricity generating station of the Urban Electric Supply Co. at the top of East Hill, Tuckingmill. Left to right: Tyrie's pumping engine house and headgear, Martin's East shaft headgear, North stamps and dressing floors, South Tincroft indoor stamps engine, then follows the 70in engine house on Harvey's shaft, partially shrouded by the first of a group of arsenic stacks. To the right of the stack, producing light-coloured smoke, is the engine house of Old Crusher whim and the next house contains the beam engine which drove the man engine and later hoisted from Downright shaft. The final headgears to be seen are on Old Tincroft and Downright shafts. To the right-hand side of the print is Bartle's Foundry. On the skyline is the top of the headgear and stack at the Marriott's shaft of the South Frances section of the Bassett mines. (1901)

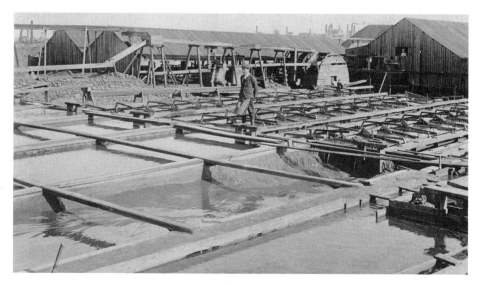

A view of a part of the dressing floors at Tincroft. It portrays a typical old-fashioned scene with settling pits and rag frames. On the skyline are the headgears, engine houses and stacks of East Pool mine. (1890s)

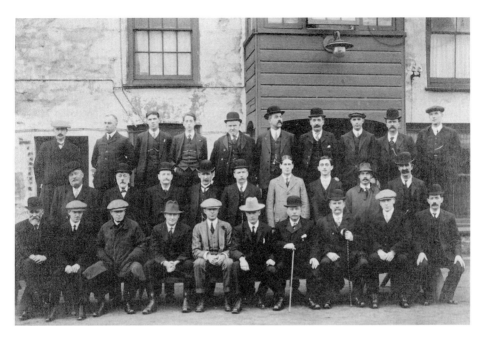

Senior management and shareholders outside the counthouse. The figure in the back row, fourth from the right, is John (Johnny) Ellis, the engineer. At a time when there were many highly regarded men in mechanical engineering in Cornwall, he was outstanding. (Early 1900s)

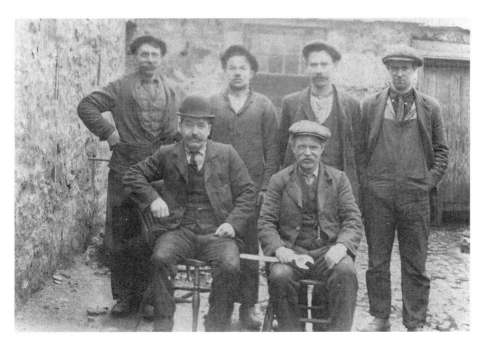

The figure in the bowler hat is Johnny Ellis, engineer of the mine, with some of the fitters from the workshops. (Early 1900s)

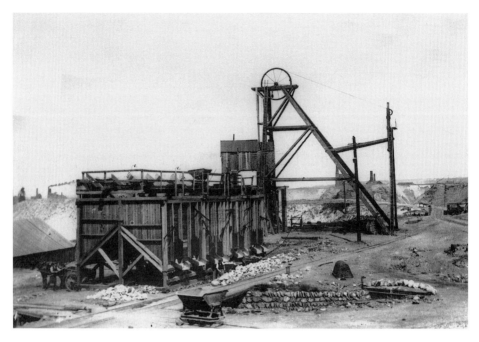

Willoughby's shaft showing the headgear, ore bins and tramways. (17 August 1915)

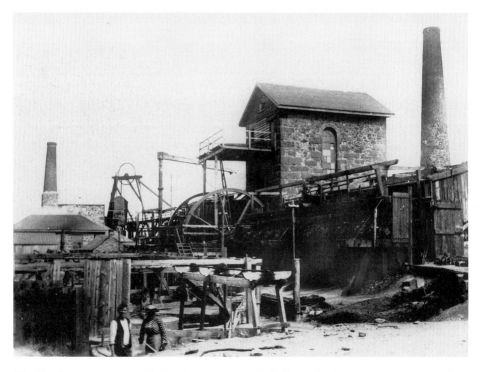

The North stamps and, in the background, Tyrie's shaft. In the foreground are two workers at the stamps floors, one of whom is a bal maiden. (17 August 1915)

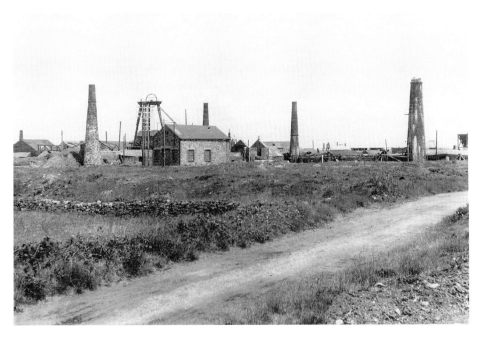

A photograph taken when the mine had ceased working. The building left of centre is the electricity substation and the headgear is on Old Tincroft shaft. Beyond the headgear is the roof of the North stamps. In the background, past the right-hand edge of the substation, is Willoughby's shaft headgear. (1921)

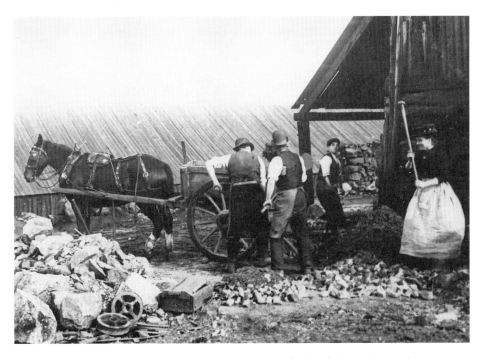

A cart being loaded by the men whilst a bal maiden spalls (breaks) ore. (c. 1900)

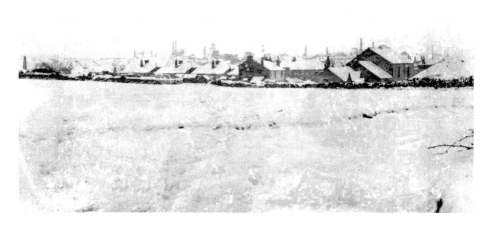

A view taken from the north-east on a snowy day, which nevertheless indicates the industrial landscape created by the mining operations. (1911)

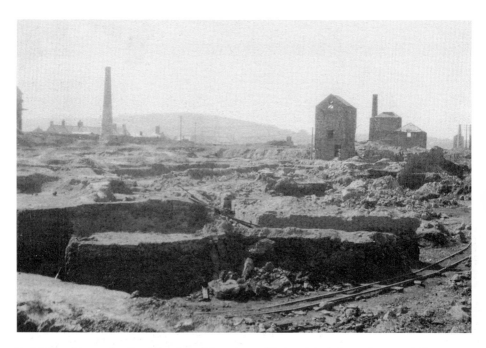

In 1923 a firm under the title of the Kingsdown Mining Co. undertook to partially reopen the mine. They proposed to work the low-grade tin, arsenic and wolfram lodes that were still standing above water level. In fact, they could work to a depth of approximately 150 fathoms as the neighbouring South Crofty mine had to hold the water to this depth in order to work their own mine. It was also decided to treat the extensive dumps and this scene shows a part of the site when this work was in progress. (1925)

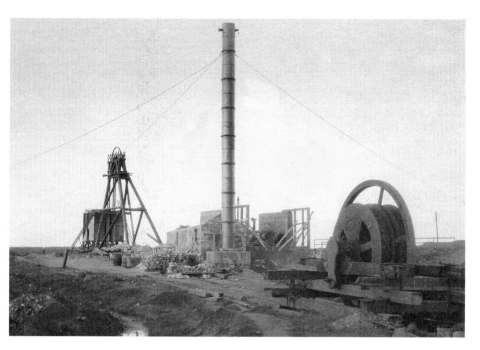

Willoughby's shaft. When the Kingsdown company commenced work at Tincroft their underground operations were centred on this shaft. The winding engine which had previously hoisted here also wound from another shaft in the former working. It was decided to move this engine to a more convenient position by Willoughby's shaft. The consulting engineer, Edgar Trestrail, decided that in order to avoid dismantling and reassembling the main shaft, cranks and winding drums could be moved as one unit by using the flywheel somewhat like a monocycle! In this manner the piece of machinery was, by the use of jacks, relocated to the new house being erected for it in the background. (1923)

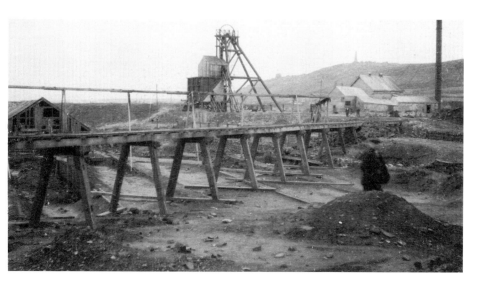

Willoughby's shaft. The plant is in full commission and the low trestle carries a tramway from the ore bins at the shaft to the North stamps. (1923)

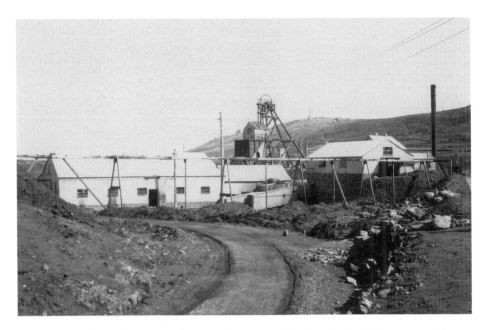

Willoughby's shaft. A later view showing the separate dressing plant which was erected near this shaft by the mill manager, Sidney Furze, to treat the old mine dumps. (1924)

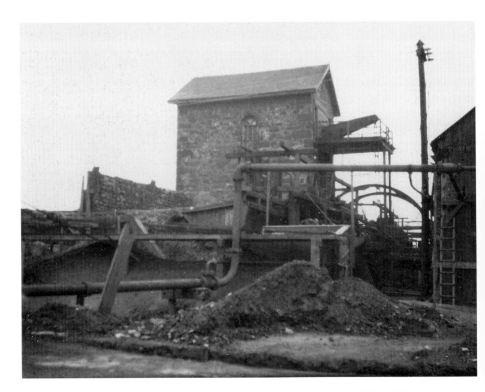

The North stamps. As far as can be ascertained, this was the last set of Cornish steam stamps to work anywhere in the world. (1925)

Harvey's 70in pumping engine. This engine, built by Harvey & Co. of Hayle in 1890, was erected to replace a 50in engine which had previously worked on this shaft called Old Sump. The shaft was thereafter renamed Harvey's in honour of the new pumping engine. This photograph shows the right-hand part of the gearwork. (1924) *(Photograph: W.A. Michell)*

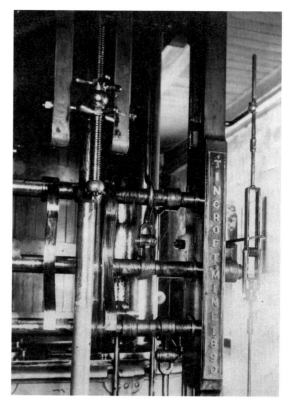

Harvey's 70in pumping engine. A further photograph which, if viewed in conjunction with the previous scene, provides a complete picture of the gearwork. (1924) *(Photograph: W.A. Michell)*

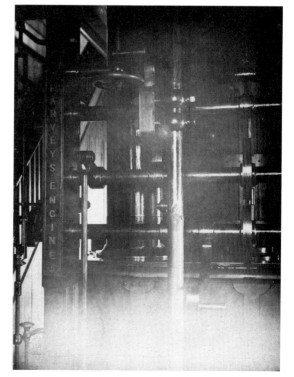

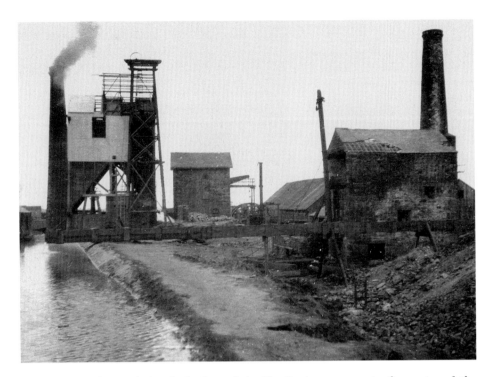

On the left is the stack for the boilers of the North stamps, seen in the centre of the photograph. The tall structure is the crusher station for the ore trammed from Willoughby's shaft before being fed into the stamps. The derelict house on the right once contained a 40in pumping engine working on Tyrie's shaft. (1924)

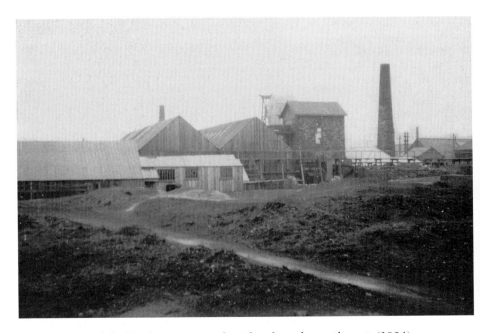

A further view of the North stamps complex taken from the north-west. (1924)

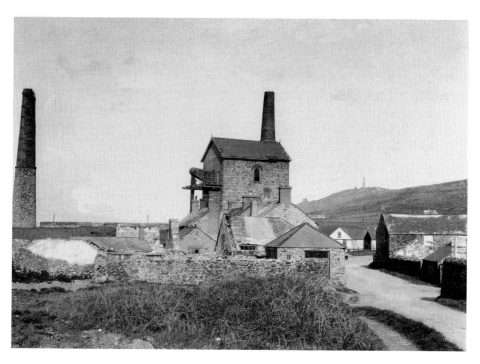

Harvey's 70in pumping engine lying derelict after the Kingsdown Company ceased operations in 1926. It had been used by them merely to pump water for mineral dressing purposes. The stack on the left served the boilers of the former 50in pumping engine which worked on this shaft. (c. 1929)

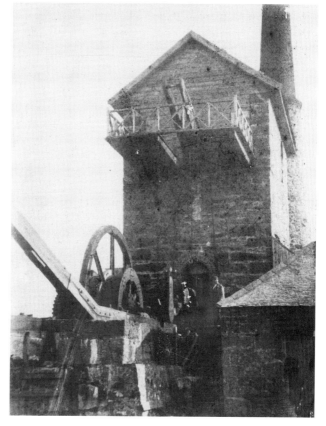

The man engine whim being scrapped. (Early 1920s)

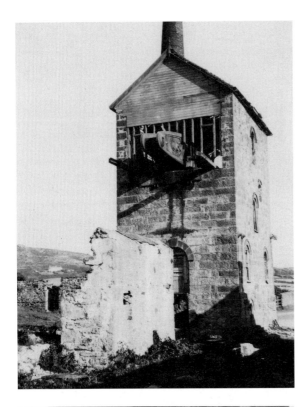

Harvey's 70in pumping engine. Taken early in the Second World War, at the time the engine was being scrapped. (*c.* 1940)

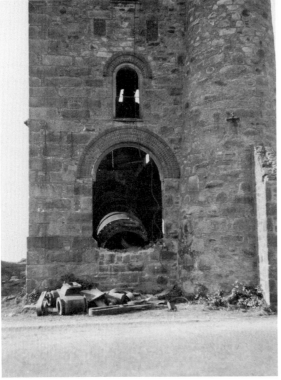

Harvey's 70in pumping engine. The partially smashed cylinder can be seen inside the house with other parts of the engine lying on the ground. (*c.* 1940)

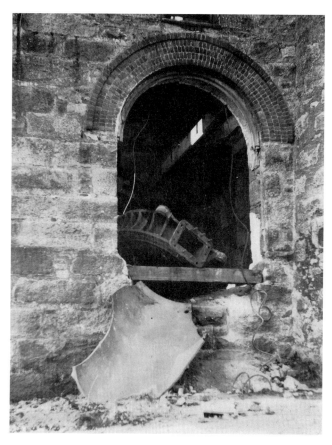

Harvey's 70in pumping engine. The cylinder bottom is inside the engine house whilst part of the smashed cylinder lies below the cylinder opening. Due to the proximity of houses, a busy road and the main railway line, the use of dynamite in this demolition was out of the question. (*c.* 1940)

Harvey's 70in pumping engine. A side view of the bob after it had been hauled out of the house. The man on the right of the photograph is the late Arthur Jory, a well-known pitman, i.e. an expert in the function, erection and maintenance of Cornish pumping equipment. (*c.* 1940)

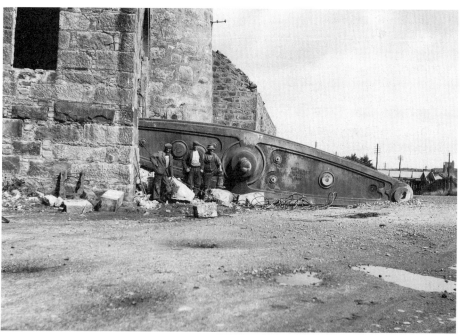

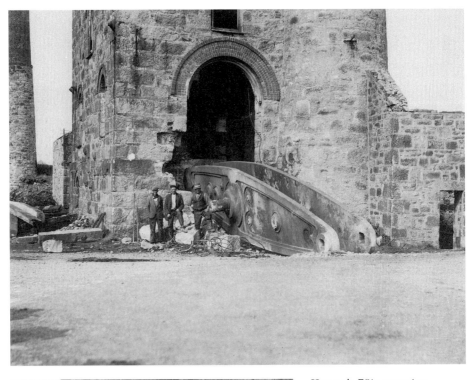

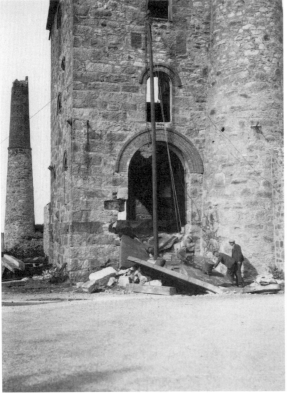

Harvey's 70in pumping engine. This scene shows clearly how, when in its working position on the bob wall, the bob was much longer indoors than out of doors, i.e. it was levered. The engine had a 10ft stroke in the cylinder and 8ft in the shaft. Arthur Jory again appears on the right of this print. (*c.* 1940)

Harvey's 70in pumping engine. The gudgeon has been burnt through with an acetylene torch and a heavy weight is being used to smash the side of the beam. Much derelict mining machinery such as this was scrapped in the Second World War due to demand for metals in the war effort. (*c.* 1940)

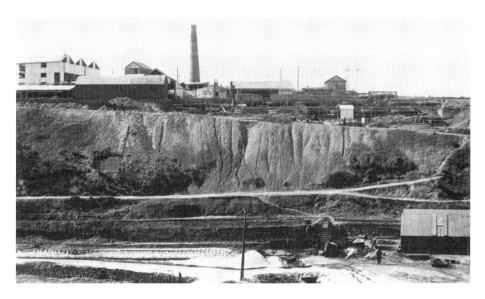

The Cornwall Tailings Co. in the Barncoose Valley. This shows the treatment works erected at that time to process the massive sand (or tailings) dumps of the Tincroft and Carn Brea mines to recover their tin content. This was formerly the site of the Wheal Agar stamps, the engine house and stack of which are in the background. (*c.* 1911)

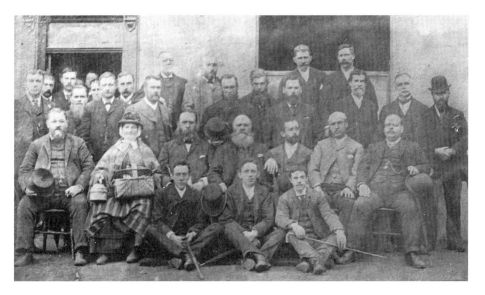

A group photograph of some of the members of the Mining Association of Cornwall & Devon (or a similar body) on the occasion of a summer outing. In the back row, the short man third from right is J.H. Collins, the famous geologist and mining author. At the extreme left of the front row is Capt Charles Bishop, manager of both East Pool mine and Wheal Grenville, together with his wife. Next to her is Capt Josiah Thomas, manager of Dolcoath mine. Fifth from left is Capt William (Willy) Thomas of Perranporth and at the end of the row is Jimmy Trevethan, purser of the Carn Brea mines. There are a number of notable persons of the period in this scene who cannot be identified. (1890s)

Carn Brea Mine

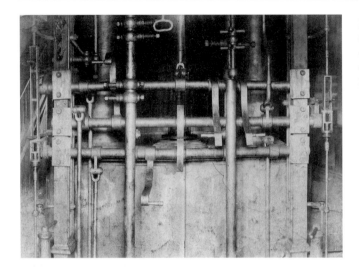

Highburrow East shaft. The gear work of the 90in pumping engine when lying idle around 1921. *(Photograph: W.A. Michell)*

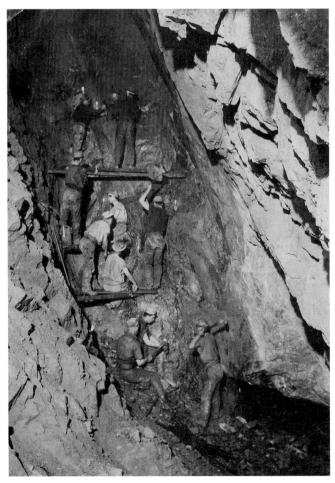

Carn Brea mine, taken at the 226 fathom level. A 'pare' of men breast stoping in a fine wide lode. (1890s) *(Photograph: J.C. Burrow FRPS)*

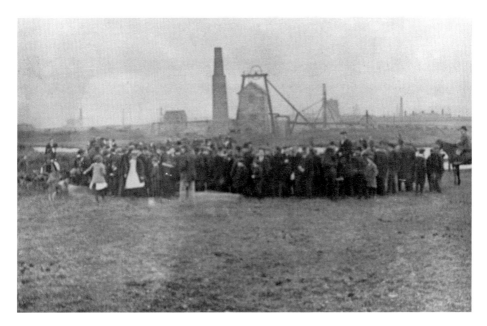

At the 'kill' with the Four Burrow Hunt, December 1904. However, the scene in the background is of some interest. Left to right: the beam winding engine on Highburrow West shaft, the stack, engine house and headgear on Old Sump shaft and, dimly in the background, the pumping engine house on Highburrow East shaft.

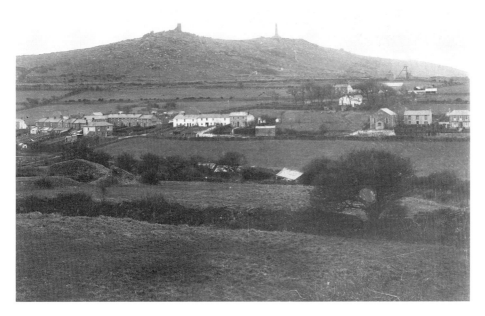

Macdonald's shaft. This was the most easterly shaft of the mine and is visible on the right of the picture. Carn Brea village forms the centre ground. (c. 1910)

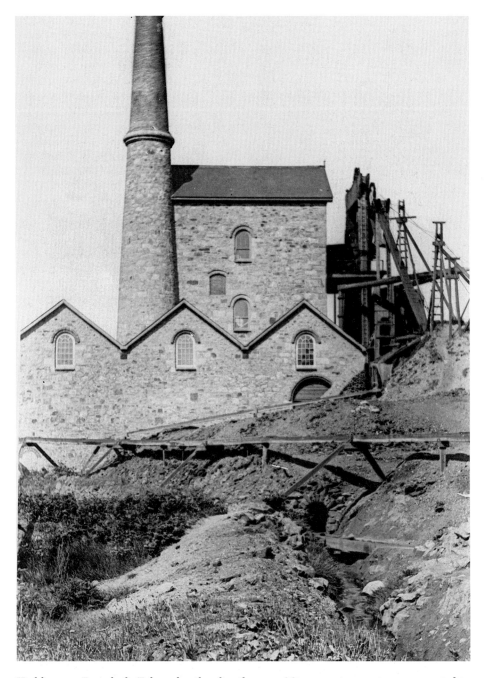

Highburrow East shaft. Taken shortly after the new 90in pumping engine was erected in 1892. This engine was steamed by four Cornish boilers contained in the building nearer to the camera. The arched drain was for use when blowing down (cleansing) the boilers. (c. 1894)

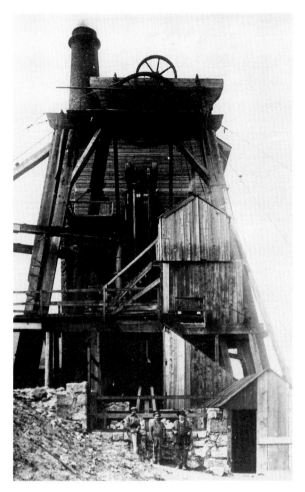

Left: Highburrow East shaft. A view from the south showing the bob of the 90in pumping engine, the capstan shears and headgear, the landing brace and a part of the surface balance box. (1890s)

Below: The hill from which the mine gets its name forms the backdrop to this scene. On the left is Old Sump shaft headgear and the 80in pumping engine house. This old stamps engine, which was one of the Sims type single-acting rotary engines, had a large amount of balance on the outdoor part of the beam. The ramshackle dressing sheds, which were not uncommon in those days, are extensive. In the foreground are long settling strips which caught the slimes still containing fine tin. In the right-hand corner of the photograph can be seen the rails of the inclined tramway which carried the ore from Highburrow East and Highburrow West shafts to the stamps. (*c.* 1900)

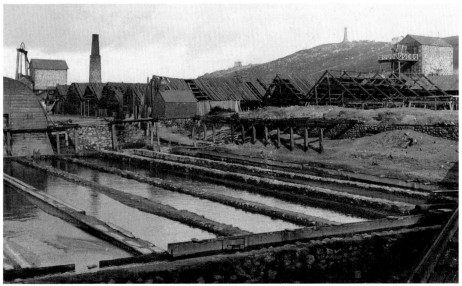

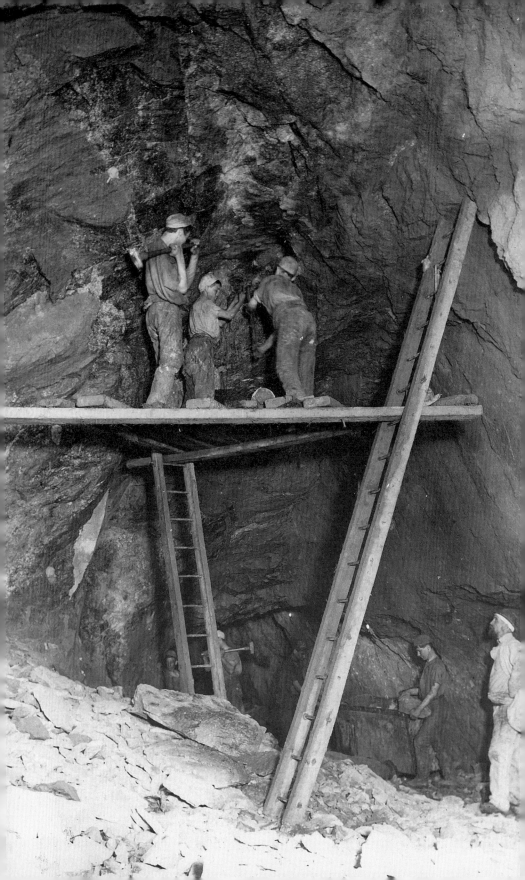

Opposite: Taken at the 274 fathom level. A triangular staging composed of ladders and planks from which a hand-labour team are drilling their holes. (1890s) *(Photograph: J.C. Burrow FRPS)*

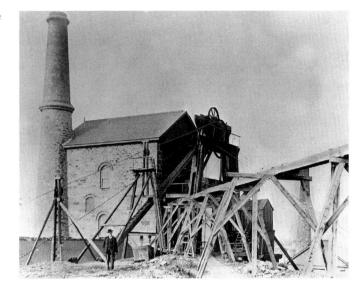

Right top: Highburrow East shaft. A view from the south-west in the early twentieth century. The building in the background is the steam capstan house and the stack served the boiler for the capstan. The capstan had originally been built at 180 degrees to this position and drew its steam from the pumping engine boilers. When it was moved it had to be provided with a separate boiler and stack. *(Photograph courtesy of The Science Museum)*

Right bottom: Highburrow East shaft showing the brickwork top of the stack of the former 90in pumping engine being removed for reuse of the bricks at the Tolvaddon stamps of East Pool & Agar mine. The steeplejack was called Jansen.(Late 1920s)

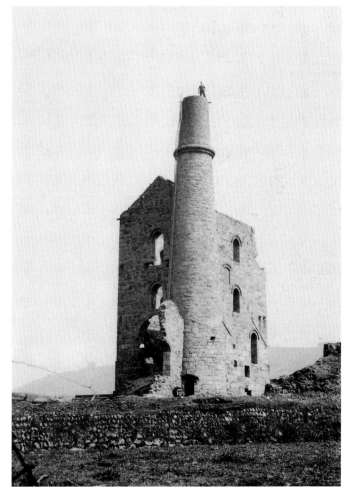

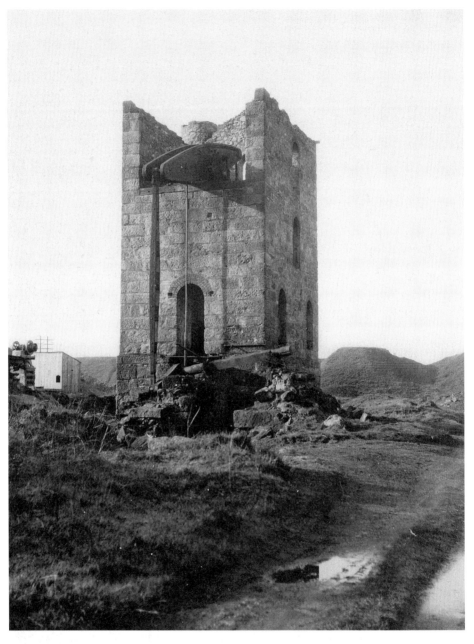

Highburrow East shaft. The beam winding engine which formerly hoisted at this shaft is being scrapped by Shapcott's metal merchants of Camborne. (Early 1930s)

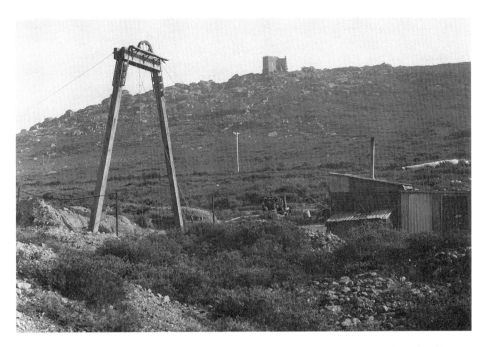

Druid's whim shaft at the time the South Crofty mine had erected their portable adit plant on this shaft as part of their extensive refurbishment of adits in the 1950s. This particular work was on the Barncoose mine adit.

Barncoose Mine

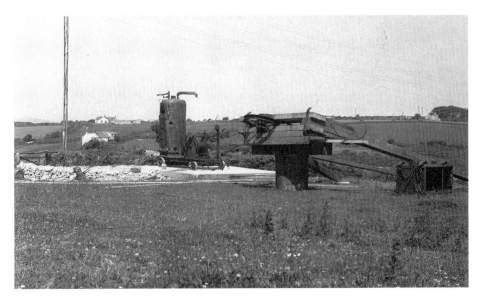

Dunn's shaft on the mine adit. The South Crofty mine portable plant is being erected in connection with clearance work in the 1950s. The Barncoose mine was absorbed into the greater Carn Brea mines when they were formed.

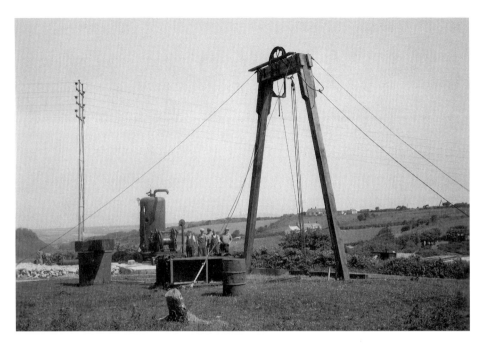

Barncoose mine adit. A further view at Dunn's shaft showing the headgear erected and the portable steam winch. (July 1955)

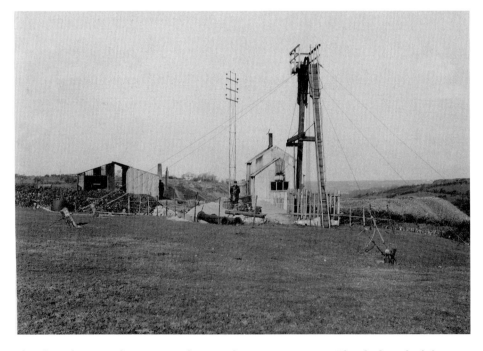

The plant shown in the previous photograph now in operation. The shed on the left was a changing room for the men. (July 1955)

Pednandrea Mine

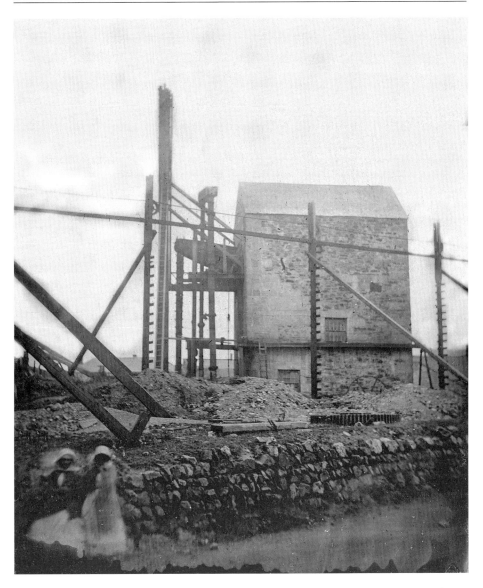

The Engine shaft. The house contained a 70in pumping engine. The pulley stands are carrying a wire rope from the winding engine to one of the many shafts further south. Note the bal maidens with their clothes blowing about on a windy day. (1880s) *(Photograph: Windmill/Michell Collection)*

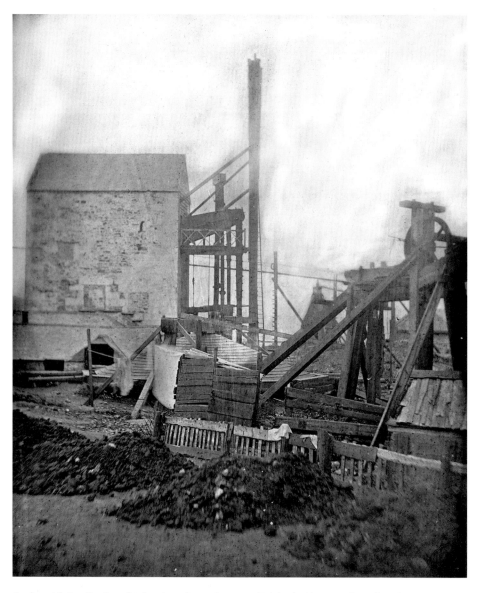

A view of the Engine shaft taken from the east. In the background, right of centre, is an end view of the headgear on Cobbler's shaft. (Late 1880s) *(Photograph: Windmill/Michell Collection)*

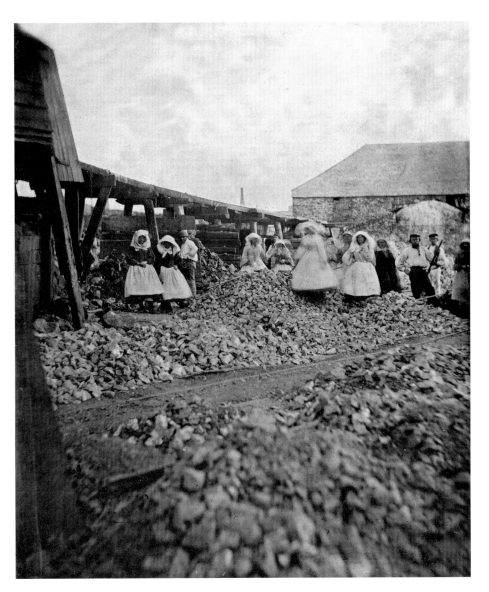

The elevated tramway carries the waggons from the shaft to the slides and spalling floors. Note the many bal maidens in the group breaking up ore. (Late 1880s) *(Photograph: Windmill/Michell Collection)*

Wheal Sparnon

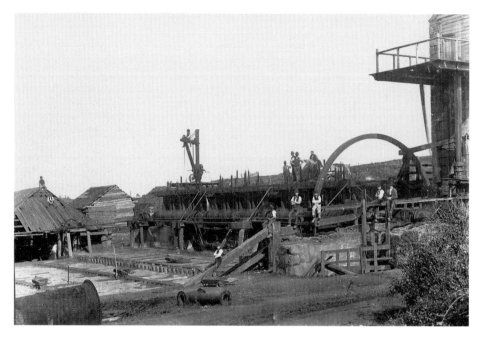

Wheal Sparnon stamps, which was used by the Pednandrea mine to crush ores for further treatment at the dressing floors on the left. (Late 1880s)

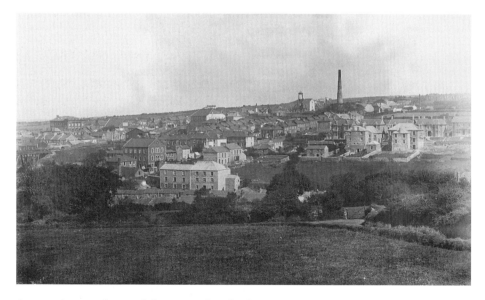

A general view of part of the town of Redruth. The very high stack of Pednandrea mine dominates the scene and to the left is the engine house, capstan shears and headgear on the Engine shaft. Further left is a stack, the stonework of which has been whitewashed. This served the boilers of the horizontal winding engine. (1880s) *(Photograph: Windmill/Michell Collection)*

On the left is the base of the lofty stack and on the right the headgear on Trevena's shaft with a group of bal maidens. The mine used a short length of the Redruth & Chasewater Railway tracks to convey ore to Wheal Sparnon stamps. As a result of this, their own mine tramway was built to the same gauge of 4ft. In this scene are horse-drawn waggons en route to the stamps. (1880s) *(Photograph: Windmill/Michell Collection)*

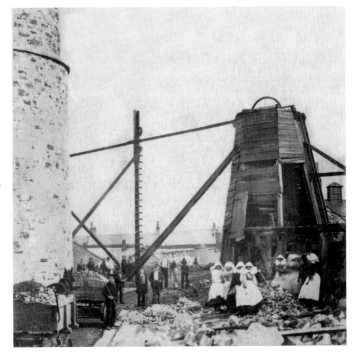

Wheal Sparnon stamps and dressing floors. To the left of the smoking stamps chimney is an arsenic stack. This scene is the present-day site of the Victoria Park at Redruth. (1880s) *(Photograph: Windmill/Michell Collection)*

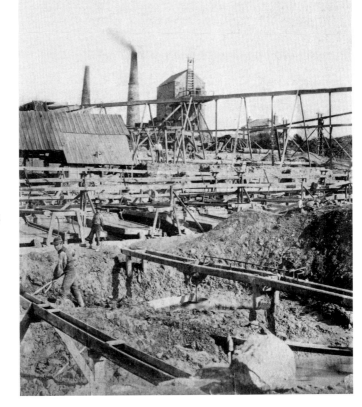

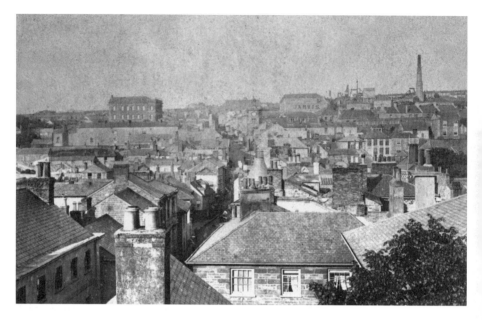

A view over the rooftops of Redruth showing the northern part of Pednandrea mine in the background. (1880s) *(Photograph: Chenhalls & Son)*

Wheal Union

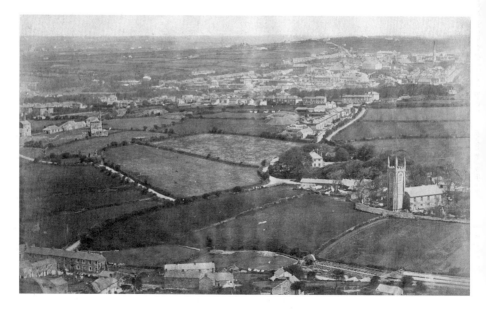

This view of Redruth was taken from the eastern end of Carn Brea. On the left is the pumping engine house on the Engine shaft of Wheal Union. To the right is a long row of buildings which contained the carpenter's shop, blacksmith's shop, dry, etc. At the end of these is the counthouse. Partly obscuring the counthouse is the beam winding engine house. Just right of centre are the burrows around the old Engine shaft. A part of Pednandrea mine is visible on the top right. (1880s) *(Photograph: Windmill/Michell Collection)*

Wheal Uny

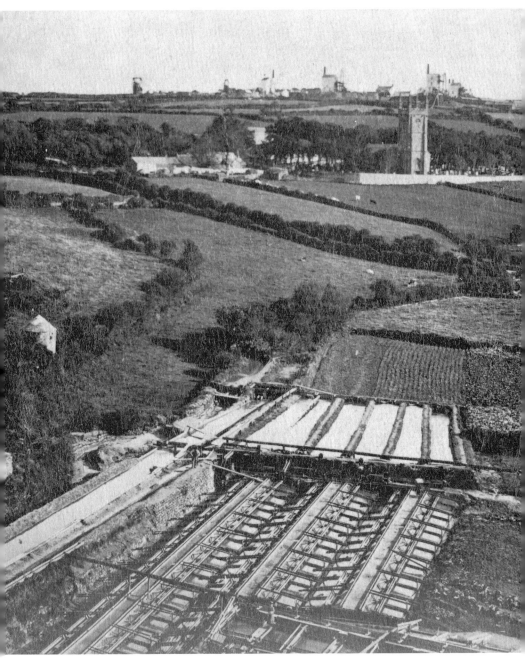

Redruth Coombe. On the skyline is Wheal Uny, showing many engine houses, etc., in the late 1880s. This is one of only two known photographs portraying the mine when working. The tower of St Uny church is prominent. In the foreground are many slime pits and rag frames in the coombe. Tin dressing such as this occupied practically the whole of the valley floor in those days. *(Photograph: Windmill/Michell Collection)*

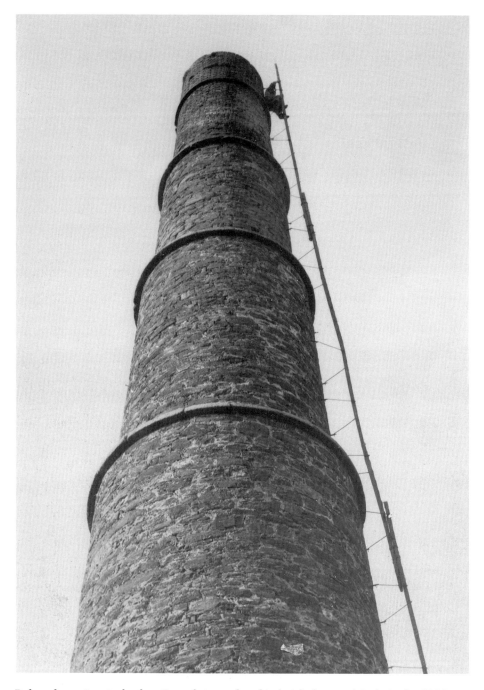

Pednandrea mine stack when it was being reduced in height by steeplejacks in the 1930s.

Opposite: Pednandrea mine stack in another view taken at the time that it was being lowered in height. It had been designed by Arthur Woolf, one of Cornwall's famous engineers. (1930s)

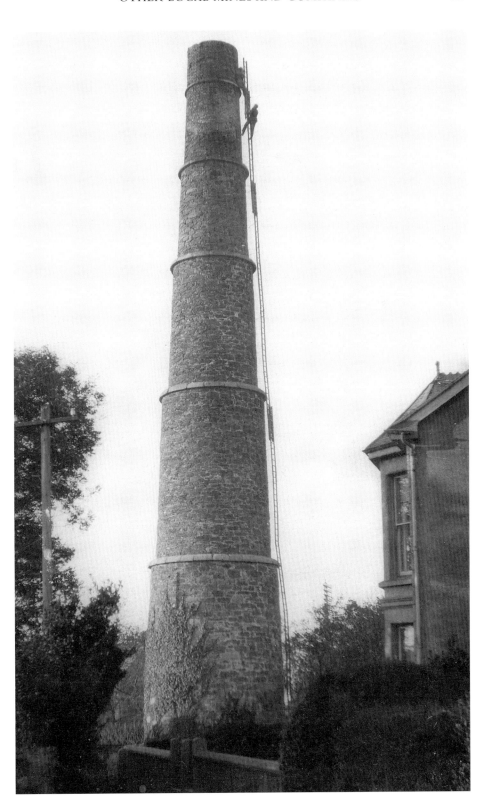

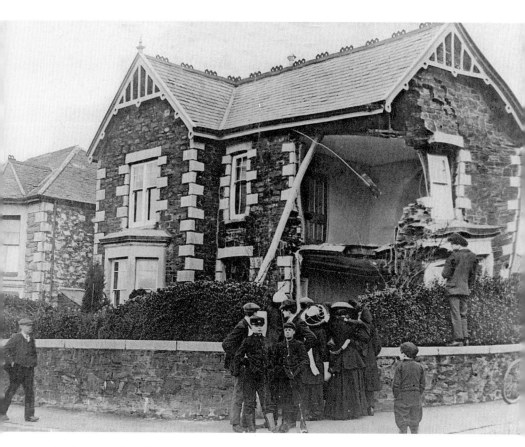

Trefusis Road, Redruth. The collapse of a house caused by the subsidence of old mine workings. (1912)

Opposite: Trefusis Road, Redruth. A further photograph taken on the following day, when the road in front of the house collapsed. This was eventually attributed to the settlement of workings on one of the Wheal Sparnon lodes. The problem was so serious that the house was demolished. Fortunately no one was injured. Wheal Sparnon adjoins Pednandrea mine. (1912)

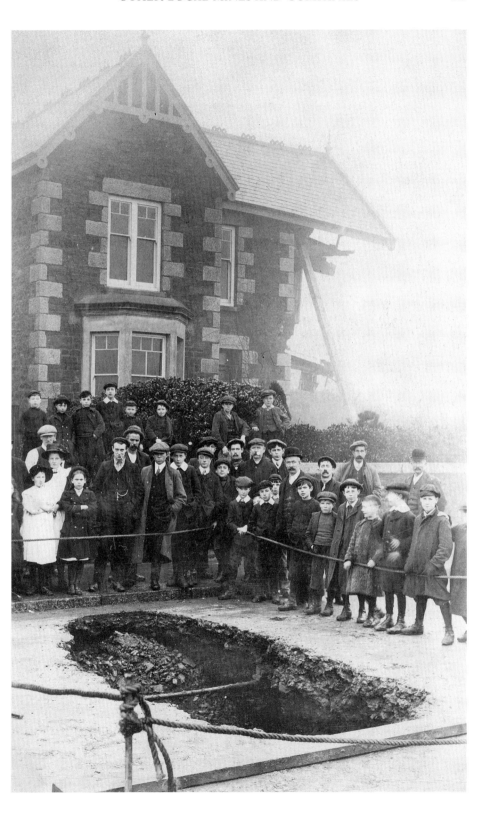

Holman Brothers

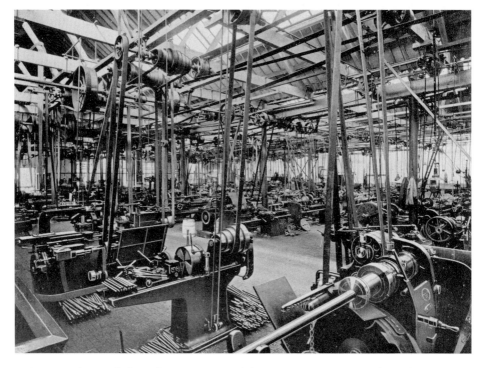

Holman Brothers Ltd of Camborne was one of the major engineering works in the county. It also survived the longest, having been in existence for nearly 170 years. Their name appears frequently in this volume and therefore it seems fitting to illustrate the works and some of their vast range of products for the mining and civil engineering industries. This scene shows their No. 3 Works machine shop in the 1920s.

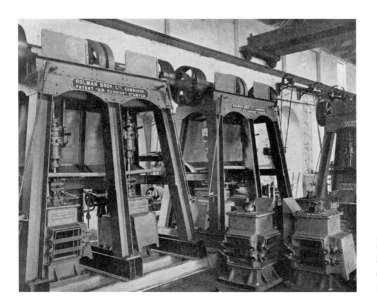

No. 1 Works showing air cushion stampers. (1920s)

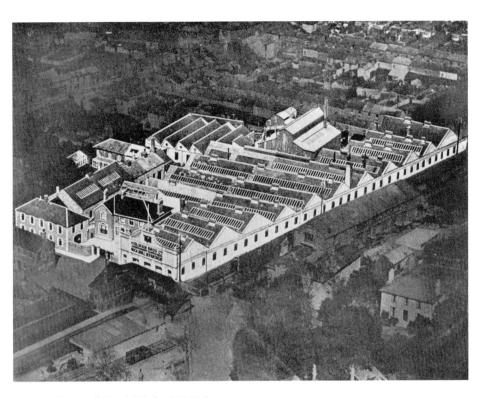

An aerial view of No. 3 Works. (1920s)

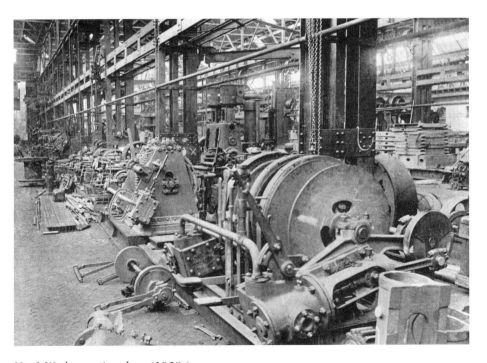

No. 1 Works erecting shop. (1920s)

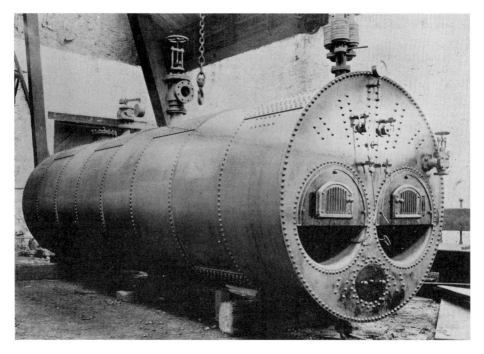

No. 2 Works (otherwise known as the boilerworks). A Lancashire-type boiler is ready for shipment. (1924)

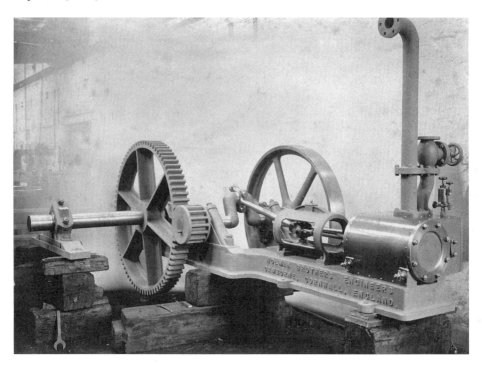

No. 1 Works erecting shop showing a 12in x 18in geared horizontal pumping engine. (1910) *(Photograph: J.C. Burrow FRPS)*